Metal forming
Mechanics
and metallurgy

Metal forming
Mechanics
and metallurgy

William F. Hosford
Professor of Metallurgical Engineering

and

Robert M. Caddell
Professor of Mechanical Engineering

University of Michigan
Ann Arbor, MI

Prentice-Hall, Inc., Englewood Cliffs, N.J. 07632

Library of Congress Cataloging in Publication Data

HOSFORD, WILLIAM F.   (date)
   Metal forming: mechanics and metallurgy.

   Includes index.
   1. Metal-work.   2. Deformations (Mechanics)   I. Caddell,
Robert M.   II. Title.
TS213.H66   671.3   82–625
ISBN   0–13–577700–3   AACR2

Editorial/production supervision
   and interior design by *Gretchen K. Chenenko*
Manufacturing buyer: *Gordon Osbourne*
Cover design by *Diane Saxe*
Cover photo courtesy of *J.W. Jones*
Art production by *Jill S. Packer*

Printed in the United States of America

10   9   8   7   6   5   4   3   2

ISBN   0-13-577700-3

Prentice-Hall International, Inc., *London*
Prentice-Hall of Australia Pty. Limited, *Sydney*
Prentice-Hall Canada Inc., *Toronto*
Prentice-Hall of India Private Limited, *New Delhi*
Prentice-Hall of Japan, Inc., *Tokyo*
Prentice-Hall of Southeast Asia Pte. Ltd., *Singapore*
Whitehall Books Limited, *Wellington, New Zealand*

Contents

8 UPPER-BOUND ANALYSIS   *143*

9 SLIP-LINE FIELD THEORY   *168*

Preface

Although metal-forming operations are diverse, the primary objective is to produce a desired shape change. Two major concerns of the engineer are the forces required for the operation and the properties of the work material. It must be recognized that material properties affect the process, and processing alters the material properties.

The major technical disciplines in metal forming are mechanics and metallurgy. Although subjects such as wear, heat transfer, and mechanical design are also important, they are not treated in this text. Of first-order importance is an understanding of the interaction of the tooling and metal during plastic deformation (mechanics) and the interrelationship of the process and the metal being processed (metallurgy). Most other texts on this subject have placed a much greater, if not a complete, emphasis on mechanics, with little concern for the metallurgical aspects involved. We feel this is a one-sided viewpoint and have included discussions of the important metallurgical aspects.

Among the topics that are either new or presented in greater detail than would be found in similarly titled texts are the following:

1. The effect of processing parameters on the inhomogeneity of the final product.
2. A comparison of plane-strain and axisymmetric operations such as extrusion and drawing.
3. A new approach to upper-bound analyses that combines the slab-force balance and a slab-energy contribution.

4. Analyses of the effects of inhomogeneity on instability.

5. A discussion of formability and a summary of properties of sheet metals.

The sequence and contents of this book are based upon a series of notes developed over the past decade for use in a senior–graduate course. Although most of our students major in Mechanical or Metallurgical Engineering, a number have come for Aerospace, Applied Mechanics, Naval Architecture, and Civil Engineering. Despite the diversity of backgrounds, the format of this text has been successful in explaining the important aspects of metal forming.

The first five chapters treat fundamental concepts. Chapters 1 through 3 are a concise review of stress, strain, yield criteria, and work hardening, and in our experience most students need such a review. Effects of strain rate and temperature plus the topic of plastic instability complete the material devoted to basic concepts.

Bulk forming processes are covered in Chapters 6 through 11. Rather than progressing from one forming process (e.g., drawing, extrusion, rolling, forging) to another, the sequence is based on different techniques of analyzing forces and patterns of flow that are common to many processes. Individual processes are treated as examples where it is most appropriate. Starting with the simplest, which is the ideal work or uniform energy balance, then progressing in sequence with the slab, upper-bound, and slip-line field analyses, the reader is provided with numerous examples that illustrate the basis, use, and limitations of each of these techniques. In Chapter 10, the influence of the geometry of the deformation zone on the structural changes that occur in the metal is discussed in detail, and Chapter 11 covers the metallurgical considerations important in bulk forming.

Chapters 12 through 16 treat many aspects of sheet forming. These have been separated from the bulk-forming processes since the primary concerns of sheet forming differ from bulk forming. The topics include bending, where springback and residual stresses are emphasized, plastic anisotropy, and the operations of cupping, redrawing, and ironing. The chapter on complex stampings treats forming limit diagrams, which have become an important basis for analyzing stamping operations. Finally, there is a discussion of the material properties important in sheet-forming operations.

Although this book is intended as a text on metal forming, we feel that it will also be of value to practicing engineers. Not only is there a review of the necessary mechanics, but some topics, such as the Δ-parameter and forming limit diagrams, should aid in solving industrial forming problems. For any conversions from English to SI units needed in this text, a table follows the preface.

This book is truly co-authored. Both of us have extensively revised the chapters written by the other. The order of authors has no significance; it was determined by the generosity of RMC instead of a coin flip as originally agreed.

Our individual interests in this subject stem from our early exposure to Walter A. Backofen (WFH) and William Johnson (RMC); and we wish to acknowledge their stimulation and encouragement. Anthony G. Atkins provided many suggestions and helpful comments over the years. We also thank William C. Leslie, J. Wayne Jones, and Suphal Agrawal for their interest and help. Among the students whose questions, opinions, and arguments forced us to greater clarity and less redundancy, we mention R. S. Raghava, A. R. Woodliff, T. Holdoe, D. J. Meuleman, and C. Vial. Finally we thank Cindy Cooper, Karen Almas, Sherry Billmeyer, and Laura Hagerman for their care and patience in typing the manuscript.

William F. Hosford
Robert M. Caddell

Ann Arbor, Michigan

USEFUL SI UNITS AND CONVERSIONS

Conversion Factors

To convert from	to	Multiply by
inch (in.)	meter (m)	2.54×10^{-2}
feet (ft)	meter	3.048×10^{-1}
pound-force (lbf)	Newton (N)	4.448
pounds/inch^2 (psi)	Pascal (Pa = N/m^2)	6.895×10^3
kilopounds/inch^2 (ksi)	Pascal	6.895×10^6
kilograms/mm^2 (kg/mm^2)	Pascal	9.807×10^6
horsepower (HP)	foot-pounds/minute	33×10^3
horsepower (HP)	watts (W)	7.457×10^2
foot-pound (ft-lbf)	Joules (J = N·m)	1.356
calories (cal)	Joules	4.187

Multiplication Factors Used in the SI System

Factor	Prefix	Symbol	Factor	Prefix	Symbol
10^9	giga	G	10^{-3}	milli	m
10^6	mega	M	10^{-6}	micro	μ
10^3	kilo	k	10^{-9}	nano	n

Metal forming
Mechanics
and metallurgy

Stress
and strain

1-1 INTRODUCTION

Since metal forming involves deformation, analyses of forming processes involve stress and strain. This chapter discusses the essential relationships used throughout this book.

1-2 STRESS

Usually, stress is defined by considering the "state of stress at a point," as shown in Fig. 1-1. The force δF acts at point P which lies on area δA. As $\delta A \rightarrow 0$, reducing the force into components that are normal and tangential to δA then defines the normal and shear stress components as

$$\sigma \equiv \frac{\delta F_n}{\delta A} \quad \text{and} \quad \tau = \frac{\delta F_t}{\delta A} \tag{1-1}$$

Since these stresses depend upon force and area, stress itself is not a vector. Figure 1-2 illustrates this point.

With the coordinate system shown, the stress σ_y acting in a direction parallel to F across area A is simply F/A. Because F has no component parallel to A, there is no shear stress acting on that plane. Now consider a plane located at angle θ which defines new coordinate axes in relation to the original x-y system. The force F has components $F_{y'}$ and $F_{x'}$ acting on the plane whose area A' equals $A/\cos\theta$. Thus, the stresses acting on the inclined plane are

$$\sigma_{y'} = \frac{F_{y'}}{A'} = \frac{F}{A}\cos^2\theta = \sigma_y\cos^2\theta \tag{1-2}$$

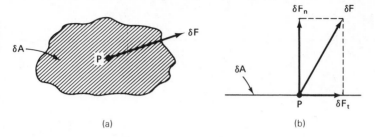

Figure 1-1 Elemental area showing total force (a) and resolved forces (b).

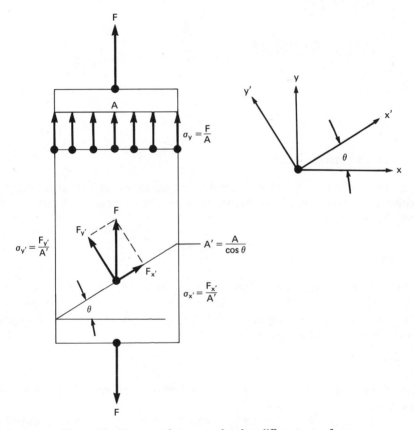

Figure 1-2 Forces and stresses related to different sets of axes.

and $$\tau_{x'} = \frac{F_{x'}}{A'} = \frac{F}{A} \sin \theta \cos \theta = \sigma_y \sin \theta \cos \theta \qquad (1\text{-}3)$$

The developments leading to Eqs. (1-2) and (1-3) have, in effect, *transformed* the stress σ_y to a new set of coordinate axes. If the point P is represented by a small body (dimensions dx, dy, dz) which is in equilibrium and is shown in

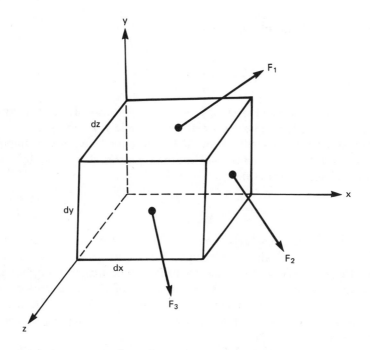

Figure 1-3 Generalized forces acting on a small body.

Fig. 1-3, then in the most general case, each face may be subjected to the total forces F_1, F_2, and F_3 as shown. Each of these forces may be resolved into components that are parallel to the three coordinate directions, and if each of these nine components is divided by the area of the face upon which it acts, the total state of stress at P is then described by the nine stress components in Fig. 1-4. This collection of stresses is called the *stress tensor*, designated as

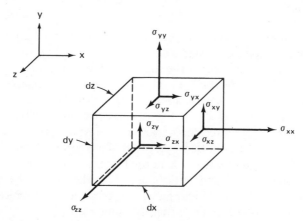

Figure 1-4 Stress element for a homogeneous state of stress. By convention, all stress components are considered positive as shown.

σ_{ij}. In tensor notation, this is expressed as,

$$\sigma_{ij} = \begin{vmatrix} \sigma_{xx} & \sigma_{yx} & \sigma_{zx} \\ \sigma_{xy} & \sigma_{yy} & \sigma_{zy} \\ \sigma_{xz} & \sigma_{yz} & \sigma_{zz} \end{vmatrix} \qquad (1\text{-}4)$$

where i and j are iterated over x, y, and z respectively. Here, two identical subscripts (e.g., σ_{xx}) indicate a normal stress, while a differing pair (e.g., σ_{xy}) indicate a shear stress. Except where the tensor nature of stresses is important, this notation will be simplified with normal stresses designated by a single subscript and shear stresses denoted by τ, so

$$\sigma_x \equiv \sigma_{xx}$$
$$\tau_{xy} \equiv \sigma_{xy} \qquad (1\text{-}5)$$

Equilibrium implies the absence of rotational effects around any axis, so $\sigma_{xy} = \sigma_{yx}$, etc., and the nine components reduce to six indepedent components. For consistency, it is necessary to introduce a convention that assigns positive and negative values to the components of the stress tensor. In this book, positive stresses are defined to act as shown in Fig. 1-4. Thus, positive normal stresses are tensile, negative normal stresses are compressive, and positive shear stresses act as shown.

The physical meaning of the double suffix notation is as follows:

1. The i subscript defines the normal to the plane upon which a component acts, while the j subscript defines the direction in which the component force acts.†
2. A combination of i and j where both are positive or both are negative defines a positive component.
3. A combination of i and j where one is positive and one is negative defines a negative component.

With the above convention, the stress σ_{xx} arises from a force acting in the positive x direction on a plane whose normal is in the positive x direction. Since both subscripts are positive, the stress component is positive and is tensile as shown. If the force acted on the same plane but in the negative x direction, the positive-negative suffix combination would indicate a negative or compressive stress. A stress such as σ_{xz} in Fig. 1-4 has two positive subscripts and is therefore positive; if the force component acted in the opposite (i.e., negative z) direction from that shown, the stress would be considered negative. Finally, if the stress state is *homogeneous*, a normal stress of the same magnitude as σ_{xx} must act upon the left-hand vertical face of the element. Such a stress would have a negative-negative combination of subscripts and, as indicated earlier, would also be defined as a positive (tensile) component.

†Use of the opposite convention should cause no confusion since $\sigma_{ij} = \sigma_{ji}$.

Example 1-1

Refer to Fig. 1-2. If a force of 5000 pounds acts as shown and the bar diameter is 2 in., determine the values of the normal and shear stresses acting upon a plane where $\theta = 30°$.

Solution

$$\sigma_y = 5000/\pi \text{ psi}$$

$$\sigma_{y'} = (5000/\pi) \cos^2 30 = 1194 \text{ psi (normal)}$$

$$\tau_{x'} = (5000/\pi) \sin 30 \cos 30 = 689 \text{ psi (shear)}$$

Note that it is assumed here that F acts uniformly across any section normal to F; thus this describes a homogeneous state of stress.

1-3 THE STRESS TENSOR

Although the components of the stress tensor are defined by Eq. (1-4), certain important physical aspects are not readily obvious from that expression. In many real situations, some of the components of the stress tensor are zero. A simple tensile test is an obvious example.

Yet even for the most general stress state there is one set of coordinate axes (1, 2, 3) along which the shear stresses vanish. The normal stresses, σ_1, σ_2, and σ_3 along these axes are principal stresses.

The magnitudes of the principal stresses are the three roots of the following cubic equation:

$$\sigma_p^3 - I_1\sigma_p^2 - I_2\sigma_p - I_3 = 0 \tag{1-6}$$

where

$$I_1 = \sigma_{xx} + \sigma_{yy} + \sigma_{zz} \tag{1-7a}$$

$$I_2 = (\sigma_{xy}^2 + \sigma_{yz}^2 + \sigma_{zx}^2 - \sigma_{xx}\sigma_{yy} - \sigma_{yy}\sigma_{zz} - \sigma_{zz}\sigma_{xx}) \tag{1-7b}$$

$$I_3 = \sigma_{xx}\sigma_{yy}\sigma_{zz} + 2\sigma_{xy}\sigma_{yz}\sigma_{zx} - \sigma_{xx}\sigma_{yz}^2 - \sigma_{yy}\sigma_{zx}^2 - \sigma_{zz}\sigma_{xy}^2 \tag{1-7c}$$

In terms of principal stresses Eqs. (1-7) become

$$I_1 = \sigma_1 + \sigma_2 + \sigma_3 \tag{1-8a}$$

$$I_2 = -(\sigma_1\sigma_2 + \sigma_2\sigma_3 + \sigma_3\sigma_1) \tag{1-8b}$$

$$I_3 = \sigma_1\sigma_2\sigma_3 \tag{1-8c}$$

For a complete development leading to Eqs. (1-6) through (1-8c) references [1,2] may be consulted. The coefficients I_1, I_2, and I_3 are independent of the coordinate system chosen in Eqs. (1-7) and (1-8), and are therefore called *invariants*. As a result, the principal stresses for a given stress state are unique.

There are two important points related to Eqs. (1-6) through (1-8). For any stress state that includes all of the shear components shown on Fig. 1-4, a determination of the three principal stresses can be made only by finding the three roots of Eq. (1-6). In addition, the invariants find importance in the development of criteria that predict the onset of yielding; this is discussed in Chap. 2.

Example 1-2

Consider a stress state where $\sigma_x = 10$, $\sigma_y = 5$, $\tau_{xy} = 3$ (all in ksi), and $\sigma_z = \tau_{zx} = \tau_{zy} = 0$. Use Eqs. (1-7a) through (1-7c) plus (1-6) to find the principal stresses in the x-y plane.

Solution. $I_1 = 15$, $I_2 = -41$, and $I_3 = 0$ from Eqs. (1-7a) etc. With Eq. (1-6),

$$\sigma_p^3 - 15\sigma_p^2 + 41\sigma_p = 0$$

or
$$\sigma_p^2 - 15\sigma_p + 41 = 0$$

The roots of this quadratic give the two principal stresses in the x-y plane. They are

$$\sigma_1 = 11.4 \text{ and } \sigma_2 = 3.6 \text{ ksi}$$

Example 1-3

Repeat Ex. 1-2 where all stresses are the same except that $\sigma_z = 8$ instead of zero.

Solution. $I_1 = 23$, $I_2 = -161$, and $I_3 = 328$. Then

$$\sigma_p^3 - 23\sigma_p^2 + 161\sigma_p - 328 = 0$$

Since $\sigma_z = 8$ is principal stress, one root of the cubic equation is known. Using $\sigma_p - 8 = 0$, this can be factored out of the cubic to give

$$\sigma_p^2 - 15\sigma_p + 41 = 0$$

which is identical to that in Ex. 1-2. Thus the remaining two principal stresses are again 11.4 and 3.6 ksi. This example shows that when z is a principal direction (whether σ_z is zero, tensile, or compressive) the remaining principal stresses are independent of σ_z.

1-4 EQUILIBRIUM EQUATIONS

If, for the model shown in Fig. 1-4, the state of stress is not homogeneous, as assumed earlier, then across each reference dimension each stress component may vary in magnitude. Assuming the body is still in equilibrium, a force balance taken in each reference direction produces the *equilibrium equations*. The detailed development is presented in the Appendix with the following results:

$$\frac{\partial \sigma_{xx}}{\partial x} + \frac{\partial \sigma_{yx}}{\partial y} + \frac{\partial \sigma_{zx}}{\partial z} = 0 \qquad (1\text{-}9a)$$

$$\frac{\partial \sigma_{xy}}{\partial x} + \frac{\partial \sigma_{yy}}{\partial y} + \frac{\partial \sigma_{zy}}{\partial z} = 0 \qquad (1\text{-}9b)$$

$$\frac{\partial \sigma_{xz}}{\partial x} + \frac{\partial \sigma_{yz}}{\partial y} + \frac{\partial \sigma_{zz}}{\partial z} = 0 \qquad (1\text{-}9c)$$

These equations are used in Chap. 9.

1-5 STRESS TRANSFORMATION EQUATIONS

A simple transformation of stresses from one coordinate system to another was discussed in connection with Fig. 1-2 and, as shown in Eqs. (1-2) and (1-3), each transformed stress is defined in terms of two angular functions. In the three-dimensional case, such a transformation is expressed as

$$\sigma_{ij} = \sum \sum \ell_{im} \ell_{jn} \sigma_{mn} \tag{1-10}$$

where m and n are iterated over the coordinate notations, x, y, and z and i and j over the new coordinate system x', y' and z'. The terms ℓ_{im} and ℓ_{jn} are the direction cosines of the angles between the axes of the two coordinate systems. The complete transformations that give $\sigma_{x'x'}, \sigma_{x'y'}$, etc., in terms of σ_{xx}, σ_{xy}, etc., and the pertinent direction cosines may be found in reference [3]. The type of problems emphasized in this book, where such transformations are required, involve simplifications of these full equations since some of the stress components are zero. Figure 1-2 illustrates such a situation; in terms of the symbolization used in Eq. (1-4), the stress $\sigma_{y'}$ in Eq. (1-2) would be $\sigma_{y'y'}$, that shown as $\tau_{x'}$ in Eq. (1-3) would be $\sigma_{y'x'}$, and the direction cosines of concern are $\ell_{y'y}$ and $\ell_{x'y}$. Thus

$$\sigma_{y'y'} = \ell_{y'y} \ell_{y'y} \sigma_{yy} = \ell_{y'y}^2 \sigma_{yy} \tag{1-11}$$

and

$$\sigma'_{xy'} = \ell_{y'y} \ell_{x'y} \sigma_{yy} \tag{1-12}$$

Since $\ell_{y'y} = \cos\theta$ and $\ell_{x'y} = \cos(90 - \theta) = \sin\theta$, Eqs. (1-11) and (1-12) are reduced to Eqs. (1-2) and (1-3).

1-6 BIAXIAL OR PLANE STRESS

If the stress components on one reference plane vanish (e.g., $\sigma_z = \tau_{zy} = \tau_{zx} = 0$) as in Fig. 1-5, a condition of biaxial or plane stress exists. Then z is a principal direction and $\sigma_z = 0$ is a principal stress. To investigate how the normal and shear stress components vary with orientation in the x-y plane, a cut is made at some arbitrary angle, ϕ, as shown in Fig. 1-5c. The stresses on this plane are σ_ϕ and τ_ϕ.

From equilibrium considerations, it can be shown that

$$\sigma_\phi = \frac{\sigma_x + \sigma_y}{2} + \frac{\sigma_x - \sigma_y}{2}\cos 2\phi + \tau_{xy}\sin 2\phi \tag{1-13}†$$

and

$$\tau_\phi = -\frac{(\sigma_x - \sigma_y)}{2}\sin 2\phi + \tau_{xy}\cos 2\phi \tag{1-14}†$$

†Two points are worth noting:
1. If specific values of σ_x, σ_y, τ_{xy} and ϕ are introduced into Eqs. (1-13) and (1-14) and the calculated values of σ_ϕ and τ_ϕ are positive, these stresses then act as shown in Fig. 1-5c; if either or both are calculated as negative stresses, they act in a direction opposite that shown in Fig. 1-5c.
2. τ_ϕ is often taken in the opposite direction shown in Fig. 1-5c; in that case, the signs in Eq. (1-14) are reversed from those shown. The comments in (1) above still apply however.

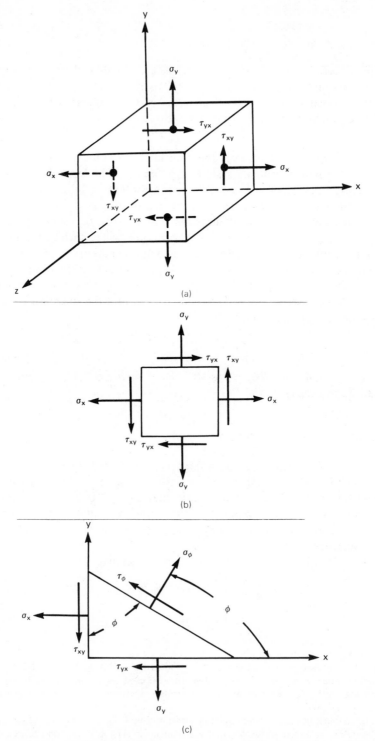

Figure 1-5 (a) Element subjected to biaxial (plane) stress. (b) Stresses as viewed in the x-y plane. (c) A cut at an arbitrary angle, ϕ.

Values of σ_ϕ on planes where τ_ϕ is zero are the two principal stresses that lie in the x-y plane. Where τ_ϕ is zero, from Eq. (1-14)

$$\tan 2\phi = \frac{2\tau_{xy}}{\sigma_x - \sigma_y} \tag{1-15}$$

and Fig. 1-6 provides the corresponding graphical relation.
Using the values of $\sin 2\phi$ and $\cos 2\phi$ from Fig. 1-6, Eq. (1-13) becomes

$$\sigma_1, \sigma_2 = \frac{1}{2}(\sigma_x + \sigma_y) \pm \frac{1}{2}[(\sigma_x - \sigma_y)^2 + 4\tau_{xy}^2]^{1/2} \tag{1-16}$$

and the values of σ_1 and σ_2 are the two remaining principal stresses (recall that $\sigma_3 = 0$ here) that act in the x-y plane.

To find the planes where τ_ϕ is a *maximum*, Eq. (1-14) is differentiated with respect to ϕ and set equal to zero. Using Fig. 1-6, the largest shear stress in the x-y plane is

$$\tau_{\max} = \frac{1}{2}[(\sigma_x - \sigma_y)^2 + 4\tau_{xy}^2]^{1/2} \tag{1-17}$$

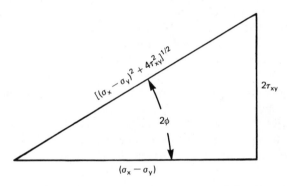

Figure 1-6 Relationship of applied stresses used to derive Eqs. (1-13) and (1-14).

Depending upon the magnitudes and signs of the applied stresses, an even larger shear stress can exist on other planes.

Example 1-4

Stresses σ_x, σ_y, and τ_{xy} are finite while σ_z, τ_{zx}, and τ_{zy} are zero. Using Eqs. (1-7a), (1-7b), (1-7c), and (1-6), show that Eq. (1-6) degenerates to Eq. (1-16).

Solution

$$I_1 = \sigma_x + \sigma_y, \quad I_2 = \tau_{xy}^2 - \sigma_x\sigma_y, \quad I_3 = 0$$

Then

$$\sigma_p^3 - (\sigma_x + \sigma_y)\sigma_p^2 - (\tau_{xy}^2 - \sigma_x\sigma_y)\sigma_p = 0$$

or

$$\sigma_p^2 - (\sigma_x + \sigma_y)\sigma_p - (\tau_{xy}^2 - \sigma_x\sigma_y) = 0$$

This quadratic gives two roots as follows:

$$\sigma_p = \frac{1}{2}(\sigma_x + \sigma_y) \pm \frac{1}{2}[(\sigma_x + \sigma_y)^2 + 4(\tau_{xy}^2 - \sigma_x\sigma_y)]^{1/2}$$

or
$$\sigma_p = \frac{1}{2}(\sigma_x + \sigma_y) \pm \frac{1}{2}[(\sigma_x - \sigma_y)^2 + 4\tau_{xy}^2]^{1/2}$$

which is Eq. (1-16).

1-7 MOHR'S CIRCLE FOR STRESS TRANSFORMATION

When one principal stress (whether tensile, compressive, or zero) is known at the outset, one principal direction is defined. In such situations, a circle plot due to Mohr[†] may be used to determine the stresses acting in the plane normal to the known principal direction. The biaxial situation discussed in Sec. 1-6 fulfills this requirement, but even if $\sigma_z = \sigma_3$ in that section were not zero, a plot of Mohr's circle can still be used with regard to the x-y plane. It is *only* essential that the axis normal to the plane of interest is a principal direction.

If this condition exists, the invariants of Eq. (1-7) can be determined and introduced into Eq. (1-6); a quadratic equation results and is identical to Eq. (1-16). The two unknown principal stresses or roots of Eq. (1-6) can be immediately determined. The major use of a Mohr's circle plot is to provide a picture of the stresses on *any* plane rotated with respect to the original x-y coordinate system. By rearranging Eqs. (1-13) and (1-14), squaring, and adding, an equation of a circle results. Details may be found in reference [2]. The circle describes a graphical tensor transformation and any components plotted in this way must be tensor quantities.[‡]

To describe both the correct magnitudes *and* the orientations of either planes or directions as one proceeds around the circle, a consistent convention is needed. The one used here is termed the *double-angle method* and entails the following:

1. Normal stresses are plotted to scale along the abscissa where tensile is positive and compressive is negative.

2. Shear stresses are plotted along the ordinate. A shear stress that induces clockwise rotation of the stress element is plotted as if it were positive, while one causing counterclockwise rotation is plotted as if it were negative.
   Note: This violates the sign convention for shear stresses portrayed in Fig. 1-4. This is the only instance where this sign convention is changed.

3. Angles between two planes or directions represented on the circle plot are double the corresponding angle on the physical plane containing the stress element.

[†]O. Mohr, *Zivilingeneur* (1882), p. 113.
[‡]This becomes more evident when strains are plotted in the form of such a circle.

Figure 1-7 shows the most general case where a Mohr's circle may be used. The original stresses are shown as σ_x (tension), σ_y (compression), and τ_{xy} (shear) in Fig. 1-7a. Following the rules in (1) and (2) above, the Mohr's circle plot is shown in Fig. 1-7b. Stresses acting on an arbitrary plane at angle ϕ from the x direction in Fig. 1-7a are then determined by laying off an angle of 2ϕ from the x direction in Fig. 1-7b. The orientations of ϕ and 2ϕ are equivalent,

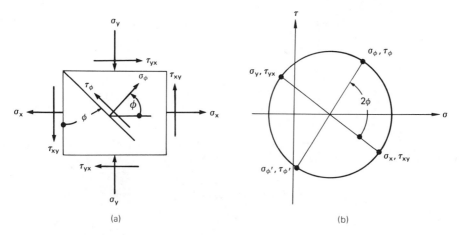

(a)   (b)

Figure 1-7   (a) Stress element and (b) Mohr's circle for a biaxial stress state.

i.e., both are counterclockwise as shown. Since all stresses are plotted to scale, the values of σ_ϕ and τ_ϕ are read directly from the circle plot. For the full element of concern, the complete stress state is given in Fig. 1-8 in terms of orien-

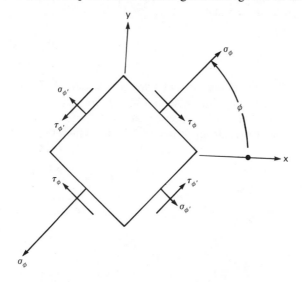

Figure 1-8   Stress element of Fig. 1-7 rotated by ϕ from the x-y axes.

tation from the original x-y coordinate system. τ_ϕ, being positive on the circle plot, must act clockwise as shown in Fig. 1-8, whereas $\tau_{\phi'}$ is counterclockwise.

Example 1-5

For the stress state in Ex. 1-2, determine the principal stresses and the largest shear stress by constructing a scaled plot of the three Mohr's circles.

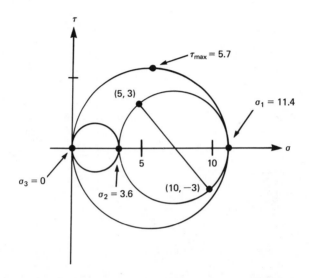

Solution

Note that $\tau_{xy} = 3$, as stated, implies counterclockwise direction in regard to Fig. 1-4, so it is plotted as a negative stress in the circle construction. Also, the largest shear stress does not act in the x-y plane.

1-8 STRAIN

As a solid is deformed, points in that body are displaced. Strain is defined in terms of such displacements but in such a way as to exclude the effects of rigid body movements like pure translation or pure rotation.

Consider first the situation shown in Fig. 1-9 where the length ℓ_o between points A and B refers to some initial condition. If under loading A moves to A',

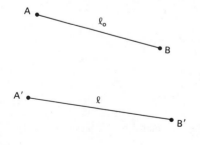

Figure 1-9 Translation, rotation, and extension of a line.

B to B', and all points between A and B move to similar relative positions between A' and B', a state of strain exists when $\ell \neq \ell_o$. Although both translation and rotation occur, it is the change in length that is used to define strain as

$$e = \frac{\ell - \ell_o}{\ell_o} = \frac{\Delta\ell}{\ell_o} \qquad (1\text{-}18)$$

where e is the *nominal* or *engineering* strain.

Note that the change in length is divided by the *original* length. For large strains, an alternative definition, proposed by Ludwik†, is more convenient. True strain, ϵ, also called *logarithmic* or *natural*, is defined such that every incremental length change is divided by the current length,

$$d\epsilon = \frac{d\ell}{\ell} \qquad (1\text{-}19)$$

which upon integrating gives

$$\epsilon = \ln\frac{\ell}{\ell_o} \qquad (1\text{-}20)$$

The following examples illustrate the convenience of true strain as it is compared with engineering strain.

Example 1-6

a) A bar of length ℓ_o is uniformly extended until its length $\ell = 2\ell_o$. Compute the values of engineering and true strain for this extension.

b) To what final length, ℓ, must a bar of initial length ℓ_o, be compressed if the strains are to be the same (except for sign) as those in part (a)?

Solution

a)
$$e = \frac{\Delta\ell}{\ell_o} = \frac{\ell_o}{\ell_o} = 1.0$$

$$\epsilon = \ln\left(\frac{\ell}{\ell_o}\right) = \ln(2) = 0.693$$

b)
$$e = -1 = \frac{\ell - \ell_o}{\ell_o}$$

so $\ell = 0$. This means the bar must be compressed to zero thickness and clearly, a strain of -1.1 is impossible to achieve.

$$\epsilon = -0.693 = \ln\left(\frac{\ell}{\ell_o}\right) \text{ so } \ell = \ell_o e^{0.693} = \frac{\ell_o}{2}$$

With this definition, the bar need only be compressed to half its initial length to induce the same true strain as in part (a). This is more physically reasonable compared with the requirements when engineering strain was used, and in Chap. 3 it will be shown that doubling the length in tension or halving it in compression will cause the same approximate degree of strengthening.

†P. Ludwik, *Elemente der Technologischen Mechanik* (Berlin: Springer, 1909).

Example 1-7

A bar of 10-in. initial length is elongated to a length of 20 in. by rolling in three stages as indicated below:

$$\text{stage 1, 10 in. increased to 12 in.}$$
$$\text{stage 2, 12 in. increased to 15 in.}$$
$$\text{stage 3, 15 in. increased to 20 in.}$$

a) Calculate the engineering strain for each and compare the *sum* of the three with the total overall value of e.

b) Repeat (a) for true strains.

Solution

a) $e_1 = \dfrac{2}{10} = 0.2,\; e_2 = \dfrac{3}{12} = 0.25,\; e_3 = \dfrac{5}{15} = 0.33,\;$ so $\; e_1 + e_2 + e_3 = 0.78\;$ but

$e_{\text{overall}} = \dfrac{10}{10} = 1.0.$

b) $\epsilon_1 = \ln\!\left(\dfrac{12}{10}\right) = 0.18,\, \epsilon_2 = \ln\!\left(\dfrac{15}{12}\right) = 0.22,\, \epsilon_3 = \ln\!\left(\dfrac{20}{15}\right) = 0.29,\;$ so $\; \epsilon_1 + \epsilon_2 + \epsilon_3$

$= 0.69$ and $\epsilon_{\text{overall}} = \ln\!\left(\dfrac{20}{10}\right) = 0.69.$

Using true strains, the sum of the increments equals the overall strain. This illustrates the *additive* property of true strains. The same is not true for engineering strains.

Example 1-8

A block of initial dimensions ℓ_o, w_o, t_o is deformed to new dimensions ℓ, w, and t.

a) Express the volume strain, $\ln(v/v_o)$, in terms of ϵ_ℓ, ϵ_w, and ϵ_t.

b) Plastic deformation causes practically no volume change. Assuming constant volume, what is the sum of the three plastic strain components?

Solution

a)
$$\frac{v}{v_o} = \frac{\ell w t}{\ell_o w_o t_o}$$

so
$$\ln\!\left(\frac{v}{v_o}\right) = \ln\!\left(\frac{\ell}{\ell_o}\right) + \ln\!\left(\frac{w}{w_o}\right) + \ln\!\left(\frac{t}{t_o}\right) = \epsilon_\ell + \epsilon_w + \epsilon_t$$

b) If $v = v_o$, $\ln\!\left(\dfrac{v}{v_o}\right) = 0$ or $\epsilon_\ell + \epsilon_w + \epsilon_t = 0.$

Note: The corresponding relations using engineering strains are not as simple.

Examples 1-6 through 1-8 show why true strains are more convenient than engineering strains because:

1. True strains for equivalent deformation in tension and compression are identical except in sign.

2. True strains are additive, the total strain being equal to the sum of the incremental strains.

3. The volume change is related to the sum of the three normal strains, and with volume constancy

$$\epsilon_x + \epsilon_y + \epsilon_z = 0 \qquad (1\text{-}21)$$

regarding the plastic portions of the total strains.

If the strains are small, then true and engineering strains are nearly equal. Expressing Eq. (1-20) as

$$\epsilon = \ln\left(\frac{\ell_o + \Delta\ell}{\ell_o}\right) = \ln\left(1 + \frac{\Delta\ell}{\ell_o}\right) = \ln(1 + e) \qquad (1\text{-}22)$$

a series expansion results in

$$\epsilon = e - \frac{e^2}{2} + \frac{e^3}{3!} + \cdots \qquad (1\text{-}23)$$

so as $e \longrightarrow o$, $\epsilon \longrightarrow e$.

Example 1-9

Calculate the ratio of the true to engineering strains, ϵ/e, for values of e of $0.001, 0.01$, $0.02, 0.05, 0.1, 0.2,$ and 0.5.

Solution

$$e = 0.001, \ \epsilon/e = \ln(1.001)/0.001 \ = 0.9995$$

$$e = 0.01, \ \ \epsilon/e = \ln(1.01)/0.01 \ \ \ = 0.995$$

$$e = 0.02, \ \ \epsilon/e = \ln(1.02)/0.02 \ \ \ = 0.990$$

$$e = 0.05, \ \ \epsilon/e = \ln(1.05)/0.05 \ \ \ = 0.976$$

$$e = 0.10, \ \ \epsilon/e = \ln(1.10)/0.10 \ \ \ = 0.953$$

$$e = 0.20, \ \ \epsilon/e = \ln(1.20)/0.20 \ \ \ = 0.911$$

$$e = 0.50, \ \ \epsilon/e = \ln(1.50)/0.50 \ \ \ = 0.811$$

This illustrates that the approximation that $\epsilon \approx e$ becomes increasingly poorer at high strains but for values of $e \leq 0.02$ it is quite acceptable. Certainly for elastic strains, where e is normally ≤ 0.01, the equivalence is obvious, and this is probably why there has been no need to invoke the definition of true strain in elasticity.

1-9 TWO DIMENSIONAL OR PLANE STRAIN FOR SMALL DISPLACEMENTS

Figure 1-10 shows a small, unstrained element, $ABDC$, the element being viewed as a point in the same way that stresses were presented previously. Assume the element is deformed into $A'B'D'C'$ where the displacements are shown as u and w.

Normal strains are related to extensions (tensile) or contractions (compressive) of the initial, unstrained length, and shear strains are defined in terms of angular distortion. The normal strain in the x direction is, by definition,

$$e_{xx} = \frac{A'C' - AC}{AC} = \frac{A'C'}{AC} - 1 \qquad (1\text{-}24)$$

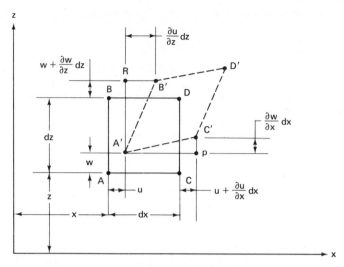

Figure 1-10   Plane strain involving small distortions.

The double subscript is used with the meaning used for stresses. With the assumption of small deformations, $A'P \approx A'C'$ and $\tan PA'C' \approx$ angle $PA'C'$. From Fig. 1-10,

$$e_{xx} \approx \frac{A'P}{AC} - 1 = \frac{dx - u + u + \dfrac{\partial u}{\partial x}\,dx}{dx} - 1 \qquad (1\text{-}25)$$

or

$$e_{xx} = \frac{\partial u}{\partial x} \qquad (1\text{-}26)$$

A similar analysis would give $e_{zz} = \partial w/\partial z$ and, if a three-dimensional situation were analyzed, $e_{yy} = \partial v/\partial y$, where v is the displacement in the y direction. References [2] and [4] discuss the three dimensional situation.

Shear strain is associated with the angular distortions shown as angles $RA'B'$ and $PA'C'$. Again, with small deformations,

$$\triangle PA'C' \approx \text{arc}\tan \frac{\dfrac{\partial w}{\partial x}\,dx}{A'P} = \text{arc}\tan \frac{\dfrac{\partial w}{\partial x}\,dx}{dx + \dfrac{\partial u}{\partial x}\,dx} \qquad (1\text{-}27)$$

or

$$\triangle PA'C' \approx \frac{\partial w}{\partial x} \quad \text{since} \quad \frac{\partial u}{\partial x} \ll 1 \qquad (1\text{-}28)$$

A similar analysis would show that $\triangle RA'B' = \dfrac{\partial u}{\partial z}$ and the total shear strain is the sum of these angles or

$$\gamma_{xz} = \frac{\partial w}{\partial x} + \frac{\partial u}{\partial z} \qquad (1\text{-}29)$$

For a three-dimensional situation, and using an interchange of subscripts, the other shear strains are

$$\gamma_{xy} = \frac{\partial u}{\partial y} + \frac{\partial v}{\partial x} \quad \text{and} \quad \gamma_{yz} = \frac{\partial v}{\partial z} + \frac{\partial w}{\partial y} \tag{1-30}$$

It is important to realize that the γ form of shear strains given in Eqs. (1-29) and (1-30) is equivalent to simple shear strain as measured in a torsion test.

1-10 THE STRAIN TENSOR

Just as the stress tensor was described by Eq. (1-4), a similar form can be used for strains as follows:

$$e_{ij} = \begin{vmatrix} e_{xx} & e_{yx} & e_{zx} \\ e_{xy} & e_{yy} & e_{zy} \\ e_{xz} & e_{yz} & e_{zz} \end{vmatrix} \tag{1-31}$$

When deformations are *small*, such as in elastic deformation, or if the incremental form $d\epsilon$ is used in Eq. (1-31), nominal or true strains are tensor quantities. With large deformations, however, the distortions causing one component of strain may affect other strain components and the analysis of strain as a tensor can lead to errors. There is another crucial point that must be understood in expressing strains as tensor quantities by Eq. (1-31). The tensor shear strain e_{xy} is equal to $\frac{1}{2}\gamma_{xy}$ and Fig. 1-11 will be helpful here. If, as in sketch (a), an

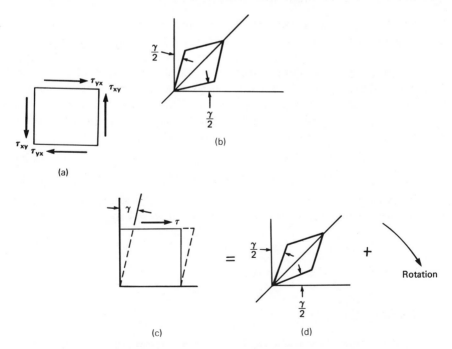

Figure 1-11 Illustration showing that pure shear, (a) and (b) is related to simple shear (c) by a rotation (d).

element is subjected to a state of pure shear, the distortion produced causes equal angular changes as shown in Fig. 1-11b. Since the total or engineering shear strain is γ, the shear strain associated with adjacent faces is $\gamma/2$ as shown. As the shear, τ_{xy}, is a tensor component, the tensor component of shear strain that parallels τ_{xy} must be $\gamma/2$ and not the total shear strain γ. Another way of indicating this is shown in Figs. 1-11c and d, where simple shear is equivalent to pure shear plus a rotation.† Thus, a tensor component, such as e_{xy} in Eq. (1-31), is *defined* as

$$e_{xy} = \frac{1}{2}\gamma_{xy} = \frac{1}{2}\left(\frac{\partial u}{\partial y} + \frac{\partial v}{\partial x}\right) \tag{1-32}$$

As with the simpler notation used for stresses in Eq. (1-5), the expressions for strains used in most of this text are as follows:

$$e_x = e_{xx}(\text{normal})$$
$$\frac{1}{2}\gamma_{xy} = e_{xy}(\text{shear}) \tag{1-33}$$

keeping in mind that any tensor shear strain is equal to one half the engineering shear strain whenever strain transformations are involved.

1-11 STRAIN TRANSFORMATION EQUATIONS

The transformation equations for small strains are identical in form to the stress equations where every normal stress, say σ_x, is replaced by its counterpart e_x and every shear stress, say τ_{xy}, is replaced by $\gamma_{xy}/2$, etc. In this text, an understanding of two-dimensional strain transformations is adequate. The description of plane strain is similar to that given in Sec. 1-6.

Assume that for a given coordinate system, $e_z = \gamma_{xz} = \gamma_{zy} = 0$, while e_x, e_y, and $\gamma_{xy} \neq 0$ describes a condition of plane strain. A typical set of transformation equations from the x-y to x'-y' coordinate axes would be

$$e_{x'} = e_x\ell_{x'x}^2 + e_y\ell_{x'y}^2 + \gamma_{xy}\ell_{x'x}\ell_{x'y} \tag{1-34}$$

and $\qquad \gamma_{x'y'} = 2e_x\ell_{x'x}\ell_{y'x} + 2e_y\ell_{x'y}\ell_{y'y} + \gamma_{xy}(\ell_{x'x}\ell_{y'y} + \ell_{y'x}\ell_{x'y}) \tag{1-35}$

Principal normal strains may be defined in terms of applied strains in a manner that produced Eq. (1-16) for principal stresses. The result is

$$e_1, e_2 = \frac{e_x + e_y}{2} \pm \frac{1}{2}[(e_x - e_y)^2 + \gamma_{xy}^2]^{1/2} \tag{1-36}$$

The maximum engineering shear strain is

$$\gamma_{\max} = [(e_x - e_y)^2 + \gamma_{xy}^2]^{1/2} \tag{1-37}$$

which parallels Eq. (1-17).

---

†If the "rotations" are subtracted from the engineering shear strains, Eq. (1-31) again results. See McClintock and Argon, *Mechanical Behavior of Materials* (Reading, Mass: Addison-Wesley, 1966), pp. 55–58, for a discussion.

If principal strains are known, strains on other planes may be found from

$$e_x, e_y = \frac{1}{2}(e_1 + e_2) \pm \frac{1}{2}(e_1 - e_2)\cos 2\theta \qquad (1\text{-}38)$$

and
$$\gamma_{xy} = (e_1 - e_2)\sin 2\theta \qquad (1\text{-}39)$$

Other relationships may be expressed for strains by using the equivalent stress equations and substituting normal strains for stresses, and the particular $\gamma/2$ for its companion value of τ.

The discussion in Sec. 1-7 for plotting a Mohr's circle of stress transformation is fully applicable here and need not be repeated. The only alteration needed is that $\gamma/2$ must be plotted along the ordinate for the reasons discussed in Sec. 1-10.

1-12 ISOTROPIC ELASTICITY

Although the emphasis of this book is on plastic deformation, a short discussion of isotropic elasticity is useful. It will provide an analogy to plastic deformation and, in a simple way, indicate why the concepts of plasticity are not a direct extension of elasticity theory.

The relation between stress and strain is called a *constitutive equation*. For elasticity, this is Hooke's law, which assumes a direct proportionality between stress and strain. For uniaxial tension, the normal strain in the loading direction is given by

$$e_1 = \frac{\sigma_1}{E} \qquad (1\text{-}40)$$

Strains in the other two principal directions are a function of Poisson's ratio, ν, and are given by

$$e_2 = e_3 = -\nu e_1 \qquad (1\text{-}41)$$

If three principal stresses are applied simultaneously, each causes Poisson strains in the other directions and the three-dimensional form of Hooke's law would show that

$$e_1 = \frac{1}{E}[\sigma_1 - \nu(\sigma_2 + \sigma_3)] \qquad (1\text{-}42)$$

while e_2 and e_3 are expressed by a proper change of subscripts. For the most generalized case involving both normal and shear stresses, Eq. (1-42) can be expressed as

$$e_x = \frac{1}{E}[\sigma_x - \nu(\sigma_y + \sigma_z)] \qquad (1\text{-}43)$$

and
$$\gamma_{xy} = \frac{\tau_{xy}}{G} = 2e_{xy} \qquad (1\text{-}44)$$

where γ_{xy} is the engineering shear strain and G the shear modulus. Expressions for e_y, γ_{xz}, etc. may then be written by interchanging subscripts in Eqs. (1-43) and (1-44).

Example 1-10

Suppose a sheet of aluminum is subjected to the stress state in Ex. 1-2. Determine the principal strains and the largest shear strain in the x-y plane
a) using appropriate equations, and
b) by plotting Mohr's circle.

Solution

a) From either Ex. 1-2 or 1-5 it was found that in the x-y plane

$$\sigma_1 = 11.4, \sigma_2 = 3.6, \text{ and } \tau_{max} \approx 4$$

Using $E = 10^4$ ksi and $v = 0.3$ for aluminum, Eq. (1-42) gives

$$e_1 = 10^{-4}[11.4 - 0.3(3.6)] = 1.03(10^{-3})$$

$$e_2 = 10^{-4}[3.6 - 0.3(11.4)] = 0.018(10^{-3})$$

$$\gamma_{12} = \tau_{12}/G$$

where from Eq. (1-51)

$$G = \frac{E}{2(1 + v)} = 3846 \text{ ksi}$$

$$\gamma_{12} = \frac{4}{3846} = 1.04(10^{-3})$$

b) Now

$$e_x = 10^{-4}[10 - 0.3(5)] = 0.85(10^{-3})$$

$$e_y = 10^{-4}[5 - 0.3(10)] = 0.2(10^{-3})$$

$$\gamma_{xy} = \frac{\tau_{xy}}{G} = \frac{3}{3846} = 0.78(10^{-3})$$

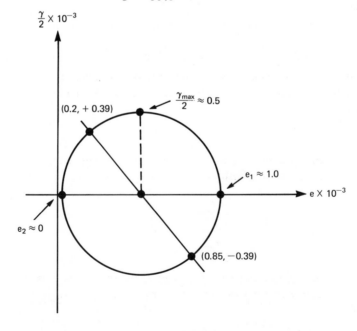

Note that $\gamma_{xy}/2 = 0.39$ and is plotted negative in combination with e_x; also, $\gamma_{max} = 2(0.5) = 1.0$, so the circle agrees with the values of e_1, e_2, and γ_{12} found in (a).

Changes in volume occur because of normal strains only, and the fractional volume change, dv/v, often called *dilatation*, is given by

$$\frac{dv}{v} = e_1 + e_2 + e_3 = \frac{(1 - 2v)}{E}(\sigma_1 + \sigma_2 + \sigma_3) \qquad (1\text{-}45)$$

using Eq. (1-42). The bulk modulus is defined as

$$B = \frac{\sigma_m}{dv/v} \qquad (1\text{-}46)$$

where σ_m is the mean normal stress expressed as

$$\sigma_m = \frac{1}{3}(\sigma_1 + \sigma_2 + \sigma_3) = \frac{1}{3}(\sigma_x + \sigma_y + \sigma_z) \qquad (1\text{-}47)$$

Using Eqs. (1-45) and (1-47), Eq. (1-46) becomes

$$B = \frac{E}{3(1 - 2v)} \qquad (1\text{-}48)$$

Under pure shear, $\tau_{max} = \sigma_1 = -\sigma_3$ and $\sigma_2 = 0$, whereas $\frac{1}{2}\gamma_{max} = e_1 = -e_3$, or $\gamma_{max} = 2e_1$. Since $\tau = G\gamma$, using Eq. (1-42) and noting that $\sigma_2 = 0$,

$$e_1 = \frac{1}{E}(\sigma_1 - v\sigma_3) = \frac{\sigma_1}{E}(1 + v) \qquad (1\text{-}49)$$

and $\qquad\qquad\qquad \dfrac{\tau_{max}}{\gamma_{max}} = G = \dfrac{\sigma_1}{2e_1} \qquad (1\text{-}50)$

Combining Eqs. (1-49) and (1-50) gives

$$G = \frac{E}{2(1 + v)} \qquad (1\text{-}51)$$

Equations (1-48) and (1-51) show that of the four elastic constants, E, G, B, and v, only two are independent.

Example 1-11

A sheet of metal is 36 in. wide, 100 in. long, and 0.030 in. thick. It is bent as shown in the figure to a 5-in. radius of curvature that is *constant* across the full 36-in. width, which implies that the strain in the 36-in. dimension is zero (i.e., plane strain). Determine the state of stress at the outer surface. Assume the deformation is elastic and no net force is *applied* in the plane of the sheet.

Solution.   Consider x to be in the length direction, y the width, and z the thickness. From Eq. (1-43)

$$e_y = 0 = \frac{1}{E}[\sigma_y - v(\sigma_x + \sigma_z)] \quad \text{and} \quad \sigma_z = 0$$

so $\qquad\qquad\qquad\qquad\qquad \sigma_y = v\sigma_x$

Neglecting any shift in the neutral axis,

$$e_x = \frac{\ell - \ell_o}{\ell_o}$$

or
$$e_x = \frac{\theta(\rho + t) - \theta\left(\rho + \frac{t}{2}\right)}{\theta\left(\rho + \frac{t}{2}\right)}$$

$$e_x = \frac{\frac{t}{2}}{\rho + \frac{t}{2}}$$

and since $\rho \gg t/2$, $e_x \approx t/2\rho$,

$$e_x = \frac{t}{2\rho} = \frac{1}{E}[\sigma_x - v(\sigma_y + \sigma_z)] = \frac{1}{E}[\sigma_x - v^2\sigma_x]$$

so
$$e_x = \frac{t}{2\rho} = \frac{\sigma_x}{E}(1 - v^2)$$

or
$$\sigma_x = \frac{t}{2\rho}\frac{E}{(1 - v^2)}, \; \sigma_y = \frac{tvE}{2\rho(1 - v^2)}$$

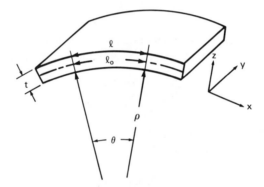

Using appropriate values for E and v, the stresses can then be found. Note that $E/(1 - v^2)$ is often called E', the plane-strain modulus. Plane strain here is a reasonable assumption because of the very large width-to-thickness ratio. At edges of the sheet, $\sigma_y = 0$, so the edges are in plane stress rather than plane strain. However, this condition exists only near the edges. If the width-to-thickness is small enough, the assumption of plane stress ($\sigma_y = 0$) throughout would be more reasonable. See Prob. 1-11.

If a metal is elastically anisotropic, E, v, and G vary with direction, so Eqs. (1-48) and (1-51) are not applicable. The reader can consult reference [3] for further details.

1-13 ELASTIC WORK OR STRAIN ENERGY

In Fig. 1-12, a bar of length x and area A is subjected to a uniaxial tensile force, F_x, which causes an incremental change in length, dx.
The incremental work, dW, is

$$dW = F_x dx \tag{1-52}$$

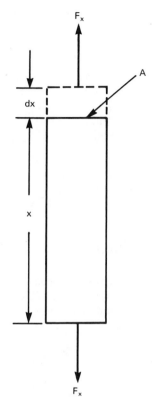

Figure 1-12 Basis for analyzing strain energy or elastic work per unit volume.

and this work per unit volume, *dw*, is

$$dw = \frac{dW}{xA} = \frac{F_x}{A}\left(\frac{dx}{x}\right) = \sigma_x \, de_x \qquad (1\text{-}53)$$

For elastic deformation, $\sigma_x = Ee_x$ for uniaxial tension; substituting in Eq. (1-53) and then integrating gives

$$w = \frac{\sigma_x e_x}{2} = \frac{Ee_x^2}{2} \qquad (1\text{-}54)$$

In the most general case, other stress components would also induce work; this leads to

$$w = \frac{1}{2}(\sigma_x e_x + \sigma_y e_y + \sigma_z e_z + \tau_{xy}\gamma_{xy} + \tau_{yz}\gamma_{yz} + \tau_{zx}\gamma_{zx}) \qquad (1\text{-}55)$$

If principal stresses are involved,

$$w = \frac{1}{2}(\sigma_1 e_1 + \sigma_2 e_2 + \sigma_3 e_3) \qquad (1\text{-}56)$$

REFERENCES

[1] W. Johnson and P. B. Mellor, *Engineering Plasticity*. New York: Van Nostrand Reinhold, 1973, pp. 44–49, 58.

[2] R. M. Caddell, *Deformation and Fracture of Solids*. Englewood Cliffs, N.J.: Prentice-Hall, 1980, pp. 5–9, 15–19, 36–37.

[3] N. H. Polakowski and E. J. Ripling, *Strength and Structure of Engineering Materials*. Englewood Cliffs, N.J.: Prentice-Hall, 1966, pp. 32–35, 116–18.

[4] H. Ford, *Advanced Mechanics of Materials*. New York: John Wiley, 1963, pp. 115–21.

APPENDIX

EQUILIBRIUM EQUATIONS

As contrasted with the homogeneous state of stress shown in Fig. 1-4, a variation of stress across the elemental dimensions of the "point model" is displayed in Fig. 1A-1, where only those components acting parallel to the *x-y* plane are shown. For the fol-

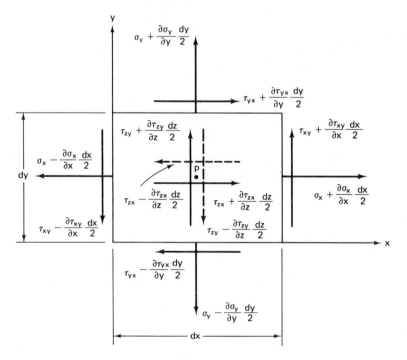

Figure 1A-1

lowing, only surface forces are considered of importance, and if the element is in equilibrium then a force balance in the x, y, and z directions must be zero.†

Considering the x direction, only the stresses σ_x, τ_{yx}, and τ_{zx} result from forces parallel to the x direction, thus

$$\Sigma\, F_x = 0 \quad \text{means} \quad \Sigma\, \text{(stress) (area)} = 0$$

so

$$\left(\sigma_x + \frac{\partial\sigma_x}{\partial x}\frac{dx}{2}\right) dy\,dz + \left(\tau_{yx} + \frac{\partial\tau_{yx}}{\partial y}\frac{dy}{2}\right) dx\,dz + \left(\tau_{zx} + \frac{\partial\tau_{zx}}{\partial z}\frac{dz}{2}\right) dx\,dy$$

$$- \left(\sigma_x - \frac{\partial\sigma_x}{\partial x}\frac{dx}{2}\right) dy\,dz - \left(\tau_{yx} - \frac{\partial\tau_{yx}}{\partial y}\frac{dy}{2}\right) dx\,dz$$

$$- \left(\tau_{zx} - \frac{\partial\tau_{zx}}{\partial z}\frac{dz}{2}\right) dx\,dy = 0 \tag{A-1}$$

Expanding and rearranging gives

$$\frac{\partial\sigma_x}{\partial x} + \frac{\partial\tau_{yx}}{\partial y} + \frac{\partial\tau_{zx}}{\partial z} = 0 \tag{A-2a}$$

and following the same approach for $\Sigma\, F_y = 0$ and $\Sigma\, F_z = 0$ gives

$$\frac{\partial\tau_{xy}}{\partial x} + \frac{\partial\sigma_y}{\partial y} + \frac{\partial\tau_{zy}}{\partial z} = 0 \tag{A-2b}$$

$$\frac{\partial\tau_{xz}}{\partial x} + \frac{\partial\tau_{yz}}{\partial y} + \frac{\partial\sigma_z}{\partial z} = 0 \tag{A-2c}$$

which are equivalent to Eqs. (1-9) in the body of the text.

PROBLEMS

1-1. For the following stress state, determine the principal stresses.

$$\sigma_{ij} = \begin{vmatrix} 10 & -3 & 4 \\ -3 & 5 & 2 \\ 4 & 2 & 7 \end{vmatrix}$$

1-2. A solid shaft of 2-in. diameter is subjected to the simultaneous loading of an axial tensile load of 20,000 lbf and a torque of 10,000 lb-in.
   (a) Using pertinent *equations* determine the principal stresses, assuming elastic behavior.
   (b) Check the answers in (a) by using a Mohr's circle construction.
   (c) Determine the magnitude of the largest shear stress.

1-3. A thin-walled tube, capped on the ends, is slowly loaded by internal pressure. During *elastic* deformation, what happens to the length of the tube, assuming end effects are negligible?

1-4. A solid 1-in. diameter rod is loaded in tension to 10,000 pounds. An identical rod is subjected to a fluid pressure of 5 ksi around its periphery, *then* subjected to the

†Note that $\Sigma\, M_x$, $\Sigma\, M_y$, and $\Sigma\, M_z$ must also be zero if the element does not rotate.

same tensile load as the first rod. Determine which rod experiences the larger shear stress (assume all deformations are elastic).

1-5. Consider a steel tube, such as the one in Prob. 1-3, whose diameter is 2 in. and wall thickness is 0.010 in. If it is loaded axially by 100 pounds in tension and exposed to an internal pressure of 30 psi, determine the magnitude of each principal strain (elastic deformations are to be assumed).

1-6. Three strain gages are attached to the surface of a component such that gage one lies along the x axis, gage two is 30° counterclockwise from one, while gage three is 90° counterclockwise from one. Under loading, the gages give the following strain readings in μ in./in.:

$$\text{one} \text{----------} 3000$$
$$\text{two} \text{----------} 3500$$
$$\text{three} \text{--------} 1000$$

(a) Find the value of γ_{xy}.
(b) Find the values of the principal strains in this plane.
(c) Using the value of γ_{xy} from (a), plot a Mohr's circle and verify the values found in (b).

1-7. A thin-walled tube of 0.01-in. wall thickness, 2-in. diameter and 3-ft. length is subjected to an internal pressure of 20 psi and an external torque of 200 lb-in. If the ends are closed,
(a) Determine the orientation of the principal stress axes in relation to the center-line direction of the tube.
(b) Calculate the principal stresses.
(c) Calculate the maximum shear stress.

1-8. Show that if

$$\tau_{\max} = \frac{1}{2}[(\sigma_x - \sigma_y)^2 + 4\tau_{xy}^2]^{1/2}$$

then

$$\tau_{\max} = \frac{1}{2}(\sigma_1 - \sigma_2)$$

1-9. A solid deforms elastically under plane strain where $e_2 = 0$. In the plane perpendicular to the "2" direction,

$$e_x = 0.005, \ e_y = 0.003, \ \text{and} \ \gamma_{xy} = 0.004$$

(a) Construct a Mohr's circle to find e_1 and e_3.
(b) Using pertinent equations, determine e_1 and e_3 as a check on part (a).

1-10. Show that true strain, defined as $\ln(\ell/\ell_o)$ for uniform deformation may be also expressed by any of the following:

$$\epsilon = \ln\left(\frac{\ell}{\ell_o}\right) = \ln\left(\frac{A_o}{A}\right) = 2\ln\left(\frac{D_o}{D}\right) = \ln\left(\frac{1}{1-r}\right)$$

where ℓ_o, A_o, and D_o are initial values of length, area, and diameter; ℓ, A, and D are instantaneous values; and r is the reduction of area defined as $r = (A_o - A)/A_o$.

1-11. A thin sheet of steel, $\frac{1}{16}$ in. thick by 10 in. wide by 36 in. long, is bent as described in Ex. 1-11. If $E = 30 \times 10^6$ psi, $\nu = 0.3$, and the neutral axis does not shift,

(a) Determine the state of stress at the outer radius, when the inner radius is 40 in., assuming the 10-in. by $\frac{1}{16}$-in. section remains rectangular and the normal strain in the 10-in. direction is zero (i.e., plane strain).

(b) What is the stress state at the very edge of the sheet where plane-stress rather than plane-strain conditions prevail?

1-12. Under plane-stress loading ($\sigma_z = 0$), measured values of e_x and e_y are 0.003 and $+0.001$ respectively. For this metal, $E = 10^7$ psi and $v = 0.3$. Find e_z.

1-13. Three mutually perpendicular normal stresses are applied to a cube of metal where $\sigma_1 = 0.30$ GPa, $\sigma_2 = 0.25$ GPa, and $\sigma_3 = -0.20$ GPa. If $E = 105$ GPa and $v = \frac{1}{3}$, find the work or strain energy per unit volume that is induced if the deformation is elastic.

1-14. A slab of metal is subjected to plane-strain deformation ($e_2 = 0$) such that $\sigma_1 = 40$ ksi and $\sigma_3 = 0$. If $E = 205$ GPa and $v = 0.33$ (note the mixed units), determine

(a) The magnitudes of the three normal strains.

(b) The strain energy per unit volume.

Macroscopic plasticity
and yield criteria

2-1 INTRODUCTION

With elastic deformation a stressed body returns to its original configuration if all loads are removed. In addition, stress and strain were related through certain elastic constants, usually E and v, via Hooke's law. Implicit there is that *any* stress will induce elastic strain. To cause plastic deformation, a particular level of stress must be reached; this has been defined as the *yield strength*. For most ductile metals, both the extent of deformation and the change in shape of the original body can continue to a large degree if the stress to cause initial yielding is continually increased. This will be discussed in detail in Chap. 3, where the use of true stress and strain proves more useful than using their nominal counterparts.

The approach to plasticity followed in this text is based upon the uniting of experimental observations with mathematical expressions in a phenomenological manner. These mathematical expressions are called *yield criteria* and their primary use is to predict if or when yielding will occur under combined stress states in terms of particular properties of the metal being stressed.

2-2 YIELD CRITERIA

Any yield criterion is a postulated mathematical expression of the states of stress that will induce yielding or the onset of plastic deformation. The most general form is

$$f(\sigma_x, \sigma_y, \sigma_z, \tau_{xy}, \tau_{yz}, \tau_{zx}) = \text{a constant, } C \tag{2-1}$$

or, in terms of principal stresses,

$$f(\sigma_1, \sigma_2, \sigma_3) = C \tag{2-2}$$

For most ductile metals that are isotropic, the following assumptions are invoked since they have been observed in many instances:

1. There is no Bauschinger† effect, thus the yield strengths in tension and compression are equivalent.
2. Constancy of volume prevails so the plastic equivalent of Poisson's ratio is 0.5.
3. The magnitude of the mean normal stress,

$$\sigma_m = \frac{\sigma_1 + \sigma_2 + \sigma_3}{3}$$

does not influence yielding.

Although we make these assumptions for the criteria to be discussed, any violations of such assumptions would require different criteria. Effects of strain rate and temperature are deferred until Chap. 5 while effects of plastic anisotropy are covered in Chap 13. It is important to realize that these constraints mean the following criteria are not universally acceptable for all solids or for all conditions under which loads are applied!

In view of assumptions (1) and (3) above, a postulated yield criterion, if plotted in three-dimensional principal stress space, must produce a prismatic surface whose cross-sectional area does not vary. This is called a yield surface. If one of the three principal stresses is held constant, which is equivalent to passing a plane through the yield surface, the resulting two-dimensional plot is called a *yield locus.*

The assumption that yielding is independent of σ_m or, as commonly called, the hydrostatic component of the total state of stress, is reasonable if plastic flow depends upon shear mechanisms such as slip or twinning. In this context, Eqs. (2-1) or (2-2) could be expressed as

$$f[(\sigma_1 - \sigma_2), (\sigma_2 - \sigma_3), (\sigma_3 - \sigma_1)] = C \tag{2-3}$$

which implies that yielding depends upon the *size* of the Mohr's circles and not their position. Figure 2-1 illustrates this point. What is indicated here is that if a stress state $(\sigma_1, \sigma_2, \sigma_3)$ will cause yielding, an equivalent stress state $(\sigma'_1, \sigma'_2, \sigma'_3)$ will also cause yielding if, for example,

$$\sigma'_1 = \sigma_1 - \sigma_m, \ \sigma'_2 = \sigma_2 - \sigma_m, \ \sigma'_3 = \sigma_3 - \sigma_m \tag{2-4}$$

†J. Bauschinger, *Civilingenieur*, 27 (1881), p. 289.

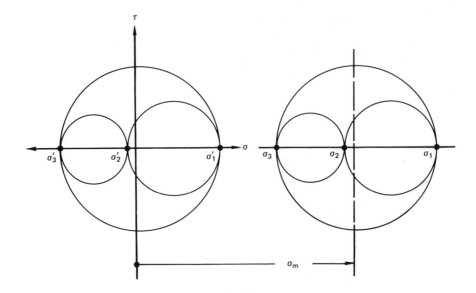

Figure 2-1 Mohr's circles showing two stress states which differ by only a hydrostatic stress, σ_m, and are, therefore, equivalent as far as yielding is concerned.

since these two states differ only by a hydrostatic stress. The stresses σ_1', etc. in Eq. (2-4) are called the deviatoric stresses since they deviate from the mean normal stress by a fixed magnitude. Thus if σ_m does not influence yielding, it must be some function of the deviatoric stresses that does lead to yielding. It is now appropriate to discuss the two most widely used yield criteria as far as ductile metals are concerned.

Example 2-1

An applied stress state is $\sigma_1 = +80$, $\sigma_2 = +45$, and $\sigma_3 = -30$ (all in MPa); yielding of the metal does not occur. If the material is then subjected to an additional fluid pressure of 50 MPa, show that yielding will still not occur.

Solution. This may be seen most readily by considering the diameters of the three Mohr's circles that result in the two cases (see Fig. 2-1).

a) Initially, $\sigma_1 - \sigma_2 = 35$, $\sigma_2 - \sigma_3 = 75$, and $\sigma_1 - \sigma_3 = 110$.

b) The fluid pressure adds a compressive stress of -50 to each of the initial stresses thus the new values become

$$\sigma_1 = 80 - 50 = 30, \sigma_2 = 45 - 50 = -5, \text{ and}$$
$$\sigma_3 = -30 - 50 = -80$$

Now the circle diameters become

$$\sigma_1 - \sigma_2 = 35, \sigma_2 - \sigma_3 = 75, \sigma_1 - \sigma_3 = 110$$

The equivalence of the three circle diameters means that the two cases are identical in regard to yielding.

2-3 TRESCA CRITERION†

This criterion postulates that yielding will occur when the largest shear stress reaches a critical value. Whenever possible, the convention $\sigma_1 > \sigma_2 > \sigma_3$ will be used, but there are cases where this relative comparison is not known a priori. In addition, this convention cannot be maintained rigorously when plots in two- or three-dimensional stress space are considered. This criterion predicts yielding when

$$\sigma_{max} - \sigma_{min} = C \quad \text{or} \quad \sigma_1 - \sigma_3 = C \quad \text{if} \quad \sigma_1 > \sigma_2 > \sigma_3 \qquad (2\text{-}5)$$

To evaluate C, a state of uniaxial tension may be used. There, $\sigma_{max} = \sigma_1, \sigma_2 = \sigma_3 = 0$, and yielding occurs when $\sigma_1 = Y$, the yield strength in uniaxial tension. Thus,

$$\sigma_1 - \sigma_3 = Y = C \qquad (2\text{-}6)$$

In the case of pure shear, $\sigma_{max} = \sigma_1, \sigma_{min} = \sigma_3 = -\sigma_1$, and $\sigma_2 = 0$. Yielding occurs when the maximum shear stress reaches the yield strength in pure shear, i.e., the shear yield strength k. At that time, $\sigma_1 = k$, so

$$\sigma_1 - \sigma_3 = 2\sigma_1 = 2k = C \qquad (2\text{-}7)$$

and the Tresca criterion becomes

$$\sigma_1 - \sigma_3 = Y = 2k \qquad (2\text{-}8)$$

Figure 2-2 shows this criterion plotted as a yield locus in two-dimensional stress space, and it is noted that this criterion is independent of the intermediate principal stress.

Example 2-2

A thin-walled tube with closed ends is to be subjected to a maximum internal pressure of 5,000 psi in service. The mean radius of the tube is to be 12 in. and it is not to yield in any region.
a) If the material has a tensile yield strength, Y, of 100,000 psi, what minimum wall thickness, t, should be specified according to the Tresca criterion?
b) If the shear yield strength, k, were specified as 40,000 psi, repeat part (a).

Solution. Since a "thin-walled" tube is indicated, the three principal stresses are $\sigma_1 = (Pr)/t$ (hoop), $\sigma_2 = (Pr)/(2t)$ (axial), and $\sigma_3 \approx 0$ (radial), where P is the pressure, r the radius, and t the wall thickness. Using Eq. (2-8),
a) $\sigma_1 = \sigma_{max'}, \quad \sigma_3 = \sigma_{min}$ so $\sigma_1 - \sigma_3 = Y$, $(5,000)12/t - 0 = 100,000$, so $t = 60,000/100,000 = 0.6$ in.
b) $\sigma_1 - \sigma_3 = 2k$, $60,000/t = 80,000$, so $t = 0.75$ in.

---

†H. Tresca, *Comptes Rendus Acad. Sci. Paris*, 59 (1864), p. 754.

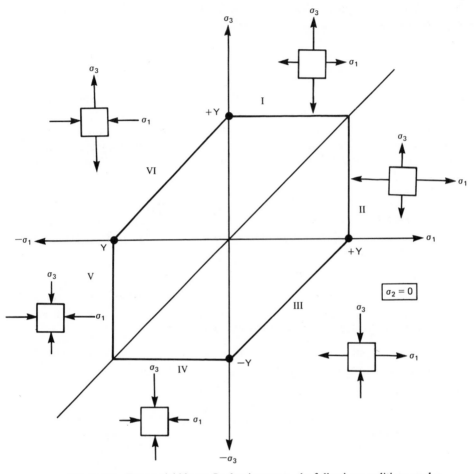

Figure 2-2 Tresca yield locus. In the six sectors, the following conditions apply:

$$
\begin{array}{ll}
\text{I} & \sigma_3 > \sigma_1 > 0, \text{ so } \sigma_3 = +Y \\
\text{II} & \sigma_1 > \sigma_3 > 0, \text{ so } \sigma_1 = +Y \\
\text{III} & \sigma_1 > 0 > \sigma_3, \text{ so } \sigma_1 - \sigma_3 = +Y \\
\text{IV} & 0 > \sigma_1 > \sigma_3, \text{ so } \sigma_3 = -Y \\
\text{V} & 0 > \sigma_3 > \sigma_1, \text{ so } \sigma_1 = -Y \\
\text{VI} & \sigma_3 > 0 > \sigma_1, \text{ so } \sigma_3 - \sigma_1 = +Y
\end{array}
$$

2-4 VON MISES CRITERION†

This postulates that yielding will occur when some value of the root-mean shear stress reaches a constant or

$$
\left[\frac{(\sigma_1 - \sigma_2)^2 + (\sigma_2 - \sigma_3)^2 + (\sigma_3 - \sigma_1)^2}{3} \right]^{1/2} = C_1 \tag{2-9}
$$

†R. von Mises, *Gött. Nach., math.-phys., Klasse*, (1913), p. 582.

Equivalently,

$$(\sigma_1 - \sigma_2)^2 + (\sigma_2 - \sigma_3)^2 + (\sigma_3 - \sigma_1)^2 = C_2 \qquad (2\text{-}10)$$

Again, uniaxial tension may be used to define C_2. Substituting $\sigma_1 = Y$ at yielding, and $\sigma_2 = \sigma_3 = 0$, the constant is $2Y^2$. For pure shear, with $\sigma_1 = k = -\sigma_3$ and $\sigma_2 = 0$, $C = 6k^2$, so the von Mises criterion is expressed as

$$(\sigma_1 - \sigma_2)^2 + (\sigma_2 - \sigma_3)^2 + (\sigma_3 - \sigma_1)^2 = 2Y^2 = 6k^2 \qquad (2\text{-}11)$$

In a more general form, this criterion can be written as

$$(\sigma_x - \sigma_y)^2 + (\sigma_y - \sigma_z)^2 + (\sigma_z - \sigma_x)^2 + 6(\tau_{xy}^2 + \tau_{yz}^2 + \tau_{zx}^2) = 2Y^2 = 6k^2$$
$$(2\text{-}12)$$

Figure 2-3 is the yield locus for this criterion while Fig. 2-4 shows the Tresca and von Mises criteria superimposed for the same value of Y. Note that the maximum differences in predictions of yielding occur along the loading paths $\alpha = -1, \frac{1}{2},$ or 2.

Several points are worth noting at this time:

1. Recalling the three invariants discussed in Chap. 1 on stress, the first invariant, I_1, equaled $(\sigma_1 + \sigma_2 + \sigma_3)$ which is $3\sigma_m$. Thus I_1 is a direct function of the hydrostatic component, and any criterion insensitive to this component in terms of yielding must be independent of I_1; both of the criteria discussed satisfy this constraint.

2. Both criteria are postulates that satisfy conditions of isotropy and independence of σ_m. Other criteria that satisfy these conditions are possible.

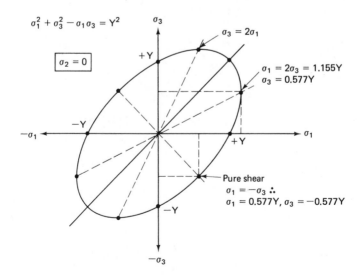

Figure 2-3   Von Mises yield locus.

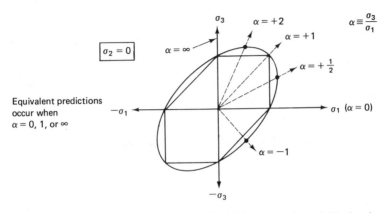

Figure 2-4 Tresca and von Mises yield loci for same value of Y, showing certain loading paths.

Figure 2-5 shows the three-dimensional plots in principal stress space of both the Tresca and the von Mises criteria, where the former is a right hexagonal prism and the latter a right circular cylinder. Both are centered on a line whose three direction cosines are equal, and any combination of stresses, $\sigma_1, \sigma_2,$ and σ_3, when added as vector components must produce a resultant that touches the yield surfaces if yielding is to occur. Figure 2-6 illustrates the resulting shapes when a plane described by $(\sigma_1 + \sigma_2 + \sigma_3) = $ constant is passed through either surface.

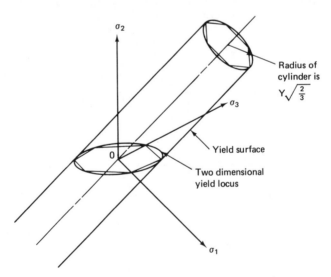

Figure 2-5 Tresca and von Mises yield surfaces in three-dimensional stress space.

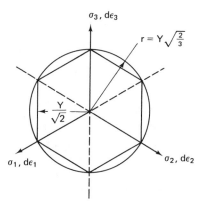

Figure 2-6 Tresca and von Mises yield surfaces projected on a plane where $\sigma_1 + \sigma_2 + \sigma_3 =$ a constant.

Example 2-3

Repeat Ex. 2-2 using the von Mises criterion.

Solution. Here $\sigma_2 = \sigma_1/2$ and $\sigma_3 = 0$, and using Eq. (2-11)

a) $(\sigma_1 - \sigma_1/2)^2 + (\sigma_1/2)^2 + (-\sigma_1)^2 = 2Y^2$ where $\sigma_1 = Pr/t$

$$\frac{2\sigma_1^2}{4} + \sigma_1^2 = \frac{6\sigma_1^2}{4} = 2Y^2, \text{ so } \sigma_1 = \frac{2}{\sqrt{3}} Y$$

$$\frac{5000(12)}{t} = \frac{2(100,000)}{\sqrt{3}}, t = 0.52 \text{ in.}$$

b) $$\frac{6\sigma_1^2}{4} = 6k^2, \text{ so } \sigma_1 = 2k$$

This is identical to part (b) of Ex. 2-2, so t again equals 0.75 in.

Note that if Y is the specified property and t is the unknown, the Tresca criterion is more conservative, but if k is the specified property both criteria predict the same value for t.

Example 2-4

Consider the same problem as in Ex. 2-3 except that t is specified as being 1 in. and the values of Y and k are unknown. Using both yield criteria,
a) Determine the value of Y to prevent yielding.
b) Determine the value of k to prevent yielding.

Solution

a) Tresca: $\qquad \sigma_1 - 0 = Y = 5000(12)/1 = 60$ ksi.

   von Mises: $\qquad \sigma_1 = \frac{2}{\sqrt{3}} Y, Y = \frac{\sqrt{3}}{2} \frac{(60,000)}{1} \cong 52$ ksi.

b) Tresca: $\qquad \sigma_1 = 2k, \text{ so } k = 60,000/2 = 30$ ksi.

   von Mises: $\qquad \frac{6\sigma_1^2}{4} = 6k^2, \text{ so } \sigma_1 = 2k = 30$ ksi.

For this particular stress state, $\sigma_1 = 2\sigma_2, \sigma_3 = 0$; this is equivalent to plane strain and the intermediate stress, $\sigma_2 = 1/2(\sigma_1 + \sigma_3) = 1/2 \, \sigma_1$. As discussed in Chap. 9, plane strain is equivalent to pure shear plus a hydrostatic component, and for such situations, if yielding is based upon k as the governing material property, both the Tresca and the von Mises criteria produce identical predictions. Finally, if Y is the governing property, the Tresca criterion is again more conservative, since it demands a higher tensile yield strength than does the von Mises criterion.

2-5 PLASTIC WORK

For the yield criteria discussed previously, the hydrostatic component of any stress state acts along the axis around which a yield surface is positioned. Due to normality†, there is no strain component of the total strain vector that acts in the direction of σ_m; as a consequence, the hydrostatic component does not tend to expand the yield surface and, in effect, does no work regardless of its magnitude. Since the deviatoric component acts in the same direction as the total strain vector, the dot product of these quantities causes maximum work as the yield surface is expanded; this plastic work is positive in the usual thermodynamic context.

A more direct approach to plastic work parallels the approach that produced Eqs. (1-52) through (1-56) in Sec. 1-13. If a bar of initial length ℓ_o is subjected to a force F acting upon an area $w_o t_o$ and an extension $d\ell$ occurs, the work done is $F d\ell$, and on a unit volume basis

$$dw = \frac{F d\ell}{w_o t_o \ell_o} = \sigma d\epsilon \qquad (2\text{-}13)$$

In the most general case, where three normal forces and three shear forces act simultaneously, the total work per unit volume is

$$dw = \sigma_x d\epsilon_x + \sigma_y d\epsilon_y + \sigma_z d\epsilon_z + \tau_{xy} d\gamma_{xy} + \tau_{yz} d\gamma_{yz} + \tau_{zx} d\gamma_{zx} \qquad (2\text{-}14)$$

In terms of principal components,

$$dw = \sigma_1 d\epsilon_1 + \sigma_2 d\epsilon_2 + \sigma_3 d\epsilon_3 \qquad (2\text{-}15)$$

2-6 EFFECTIVE STRESS

With either yield criterion, it is useful to define an effective stress denoted as $\bar{\sigma}$ which is a function of the applied stresses. If the *magnitude* of $\bar{\sigma}$ reaches a critical value, then the applied stress state will cause yielding; in essence, it has reached an effective level. For the von Mises criterion,

$$\bar{\sigma} = \frac{1}{\sqrt{2}}[(\sigma_1 - \sigma_2)^2 + (\sigma_2 - \sigma_3)^2 + (\sigma_3 - \sigma_1)^2]^{1/2} \qquad (2\text{-}16)$$

while for the Tresca criterion,

$$\bar{\sigma} = \sigma_1 - \sigma_3 \quad \text{where} \quad \sigma_1 > \sigma_2 > \sigma_3 \qquad (2\text{-}17)$$

†This is discussed in Sec. 2-9.

Note that when $\bar{\sigma} = Y$, either criterion predicts yielding, whereas $\bar{\sigma}$ must reach $\sqrt{3}\,k$ according to von Mises and $2k$ according to the Tresca criterion for yielding.

2-7 EFFECTIVE STRAIN

Effective strain is *defined* such that the incremental work per unit volume is

$$dw = \bar{\sigma}\,d\bar{\epsilon} = \sigma_1\,d\epsilon_1 + \sigma_2\,d\epsilon_2 + \sigma_3\,d\epsilon_3 \qquad (2\text{-}18)\dagger$$

For the von Mises criterion, the effective strain is given by

$$d\bar{\epsilon} = \frac{\sqrt{2}}{3}[(d\epsilon_1 - d\epsilon_2)^2 + (d\epsilon_2 - d\epsilon_3)^2 + (d\epsilon_3 - d\epsilon_1)^2]^{1/2} \qquad (2\text{-}19)$$

which may be expressed in a simpler form as

$$d\bar{\epsilon} = \left[\frac{2}{3}(d\epsilon_1^2 + d\epsilon_2^2 + d\epsilon_3^2)\right]^{1/2} \qquad (2\text{-}20)$$

In terms of *total plastic* strains,

$$\bar{\epsilon} = \left[\frac{2}{3}(\epsilon_1^2 + \epsilon_2^2 + \epsilon_3^2)\right]^{1/2} \qquad (2\text{-}21)$$

In the Appendix, Eq. (2-20) is derived; the equivalence of Eqs. (2-19) and (2-20) is also shown.

Example 2-5

In each of the following cases, plastic deformation has occurred. Using Eqs. (2-15) and (2-18), determine the plastic work per unit volume for
a) uniform uniaxial tension,
b) plane-strain compression where $\sigma_3 = 0$, and
c) plane stress where $\sigma_1 = 2\sigma_2, \sigma_3 = 0$.

Solution

a) Here, $\sigma_1 \neq 0, \sigma_2 = \sigma_3 = 0, d\epsilon_1 = -2d\epsilon_2 = -2d\epsilon_3$. With Eq. (2-15), $dw = \sigma_1 d\epsilon_1$, other terms being zero. From Eqs. (2-16), (2-18), and (2-20), $\bar{\sigma} = \sigma_1$ and $d\bar{\epsilon} = d\epsilon_1$, so $dw = \sigma_1 d\epsilon_1$ again.

b) Here, $\sigma_2 = \frac{1}{2}(\sigma_1 + \sigma_3) = \frac{1}{2}\sigma_1$ and $d\epsilon_2 = 0, d\epsilon_1 = -d\epsilon_3$

$$dw = \sigma_1 d\epsilon_1 + \frac{1}{2}\sigma_1(0) + 0 = \sigma_1 d\epsilon_1$$

$$d\bar{\epsilon} = \left[\frac{2}{3}(d\epsilon_1^2 + 0 + (-d\epsilon_1)^2)\right]^{1/2} = \frac{2}{\sqrt{3}}d\epsilon_1$$

---

†It should be noted that here, and in the subsequent use of effective strain and the flow rules, the strains ϵ_i refer to the *plastic* portion of strain. Others make this clear by using the notation ϵ_i^p, where the superscript P denotes the plastic portion of the total strain. There, $\epsilon_i^p = \epsilon_i(\text{total}) - \epsilon_i(\text{elastic})$. We have avoided this more complex notation since the major concern in this book is with large strains where elastic effects are negligible.

$$\bar{\sigma} = \frac{1}{\sqrt{2}}\left[\left(\frac{\sigma_1}{2}\right)^2 + \left(\frac{\sigma_1}{2}\right)^2 + (-\sigma_1)^2\right]^{1/2} = \frac{\sqrt{3}}{2}\sigma_1$$

$$dw = \left(\frac{\sqrt{3}}{2}\sigma_1\right)\left(\frac{2}{\sqrt{3}}d\epsilon_1\right) = \sigma_1\,d\epsilon_1 \text{ as before.}$$

c) Here, $d\epsilon_1 + d\epsilon_2 + d\epsilon_3 = 0$, $d\epsilon_3 = -d\epsilon_1 - d\epsilon_2$

$$dw = \sigma_1\,d\epsilon_1 + \frac{\sigma_1}{2}d\epsilon_2 + 0 = \sigma_1\left(d\epsilon_1 + \frac{d\epsilon_2}{2}\right)$$

$$d\bar{\epsilon} = \left(\frac{2}{3}\left[d\epsilon_1^2 + d\epsilon_2^2 + (-d\epsilon_1 - d\epsilon_2)^2\right]\right)^{1/2}$$

$$d\bar{\epsilon} = \left[\frac{2}{3}(d\epsilon_1^2 + d\epsilon_2^2 + d\epsilon_1^2 + 2d\epsilon_1\,d\epsilon_2 + d\epsilon_2^2)\right]^{1/2}$$

$$d\bar{\epsilon} = \left[\frac{4}{3}(d\epsilon_1^2 + d\epsilon_1\,d\epsilon_2 + d\epsilon_2^2)\right]^{1/2}$$

Since $\sigma_2 = \frac{1}{2}\sigma_1$, $\bar{\sigma} = (\sqrt{3}/2)\sigma_1$ as in part (b). However, since $\sigma_2 = \sigma_m$ in this case, $d\epsilon_2 = 0$, so

$$\bar{\sigma}\,d\bar{\epsilon} = \frac{\sqrt{3}}{2}\sigma_1\left[\frac{4}{3}(d\epsilon_1^2)\right]^{1/2} = \sigma_1\,d\epsilon_1$$

which equals dw above, where $d\epsilon_2 = 0$.

For the Tresca criterion, the effective strain is

$$d\bar{\epsilon} = |d\epsilon_i|_{\max} \tag{2-22}$$

where i refers to the principal directions. Thus, $\bar{\epsilon}$ for the Tresca criterion equals the absolute value of the largest principal strain. This has found little use and is not pursued further except to note that the relations

$$|\epsilon_i|_{\max} \leq \bar{\epsilon}_{\text{Mises}} \leq 1.15\,|\epsilon_i|_{\max} \tag{2-23}$$

provide a quick and simple check when evaluating $\bar{\epsilon}$ for the von Mises criterion. When $\bar{\sigma}$ is defined by Eq. (2-16), the companion value of $d\bar{\epsilon}$ or $\bar{\epsilon}$ must be defined by Eqs. (2-19), (2-20), or (2-21). If $\bar{\sigma}$ is defined by Eq. (2-17), that is the Tresca definition, then it is *not* correct to define $d\bar{\epsilon}$ by Eq. (2-19) even though this has been done on occasion. In this text, whenever both $\bar{\sigma}$ and $d\bar{\epsilon}$ are used in conjunction, Eqs. (2-16) and (2-19), (2-20), or (2-21) should be used.

It should be noted that the coefficients of $1/\sqrt{2}$ and $\sqrt{2}/3$ in Eqs. (2-16) and (2-19) result by relating both $\bar{\sigma}$ and $d\bar{\epsilon}$ to tensile deformation. This means that a true stress-true strain curve obtained under uniaxial tension is an effective stress-effective strain curve, and this point is explained in greater detail in Chap. 3.

Throughout this book it is assumed that the relation $\bar{\sigma} = K\bar{\epsilon}^n$ is unique for a given material; thus, the tensile stress-strain relation may be used to predict the stress-strain behavior under other forms of loading, although this must be viewed as a first order *approximation* in certain cases.

Although the relating of $\bar{\sigma}$ and $\bar{\epsilon}$ by this power-law expression is the main reason for defining effective stress and strain, caution must be exercised. Experiments have shown that the $\bar{\sigma}$-$\bar{\epsilon}$ behavior derived from tests along other loading paths may not coincide with the uniaxial tension results. Differences are usually greatest at large strains and may be due, at least in part, to the crystallographic textures developed during deformation.

Also, when the deformation involves substantial changes in loading paths, for example, tension in an x direction followed by compression in a y direction, use of an effective strain function is even more questionable. In the case where Eq. (2-21) is used to compute the total plastic strain during compression in the x direction followed by tension in the x direction, one might conclude that $\bar{\epsilon}$ is zero, thereby assuming that no strain hardening occurred. This would be an absurd conclusion! It can be best explained by comparing two approaches used in plasticity. One, called *total* or *deformation plasticity*, is concerned with the total deformation based upon starting and ending points, and this approach would indicate the $\bar{\epsilon}$ value of zero just mentioned. The other approach, called *incremental plasticity*, would indicate that for the deformation just mentioned, $\bar{\epsilon} = \sum \bar{\epsilon}_i = \Delta \bar{\epsilon}_{\text{comp}} + \Delta \bar{\epsilon}_{\text{tensile}}$. This is a far more realistic prediction; however, during such a complete reversal of strain paths, it may lead to an overestimation of $\bar{\epsilon}$ because of the Bauschinger effect. Also, if this reversal were repeated for a number of cycles, the full strain-hardening effect would begin to saturate and a calculation of $\bar{\epsilon}$ by incremental theory would become more in error.

2-8 FLOW RULES OR PLASTIC STRESS-STRAIN RELATIONS

For elastic deformation, discussed in Sec. 1-12, stress and strain are related by Hooke's law; for example,

$$e_1 = \frac{1}{E}[\sigma_1 - \nu(\sigma_2 + \sigma_3)] \qquad (2\text{-}24)$$

Similar relations, called *flow rules*, have been developed for plastic deformation. A simple way to introduce these is to consider uniaxial tension after initial yielding has occurred. There $\sigma_1 \neq 0, \sigma_2 = \sigma_3 = 0$, and $\sigma_m = \sigma_1/3$. The deviatoric stresses, obtained by subtracting σ_m from each principal stress, are

$$\sigma_1' = \frac{2\sigma_1}{3}, \sigma_2' = \sigma_3' = \frac{-\sigma_1}{3}$$

or
$$\sigma_1' = -2\sigma_2' = -2\sigma_3' \qquad (2\text{-}25)$$

With constancy of volume and isotropy,

$$de_1 = -2d\epsilon_2 = -2d\epsilon_3 \qquad (2\text{-}26)$$

so that

$$\frac{d\epsilon_1}{d\epsilon_2} = -2 = \frac{\sigma_1'}{\sigma_2'} \qquad (2\text{-}27)$$

and so forth. Equation (2-27) implies that

$$\frac{d\epsilon_1}{\sigma_1'} = \frac{d\epsilon_2}{\sigma_2'} = \frac{d\epsilon_3}{\sigma_3'} \qquad (2\text{-}28)$$

In Eq. (2-28), the strains are the plastic increments and at any instant of deformation, the ratio of these increments to the current deviatoric stresses is constant.

A more useful form of Eq. (2-28) is

$$de_1 = \frac{d\bar{\epsilon}}{\bar{\sigma}}\left[\sigma_1 - \frac{1}{2}(\sigma_2 + \sigma_3)\right] \tag{2-29}$$

and the similarity of Eq. (2-29) with (2-24) is noted since $\frac{1}{2}$ replaces v and the ratio $d\bar{\epsilon}/\bar{\sigma}$ replaces $1/E$. It is also important to point out that although the ratio of $d\bar{\epsilon}/\bar{\sigma}$ changes during deformation, it is always positive since both of the involved quantities are positive by their definitions.

In a general sense, the flow rules Eq. (2-29) may be derived by using

$$d\epsilon_{ij} = \frac{\partial f}{\partial \sigma_{ij}} d\lambda \tag{2-30}$$

where f is taken as the yield criterion. This is related to what is called the *plastic potential*.† If the von Mises criterion is used for f and the appropriate differentiation conducted, Eq. (2-29) results. As a final comment, the above relations can be rearranged such that the ratios of the strain increments are

$$d\epsilon_1 : d\epsilon_2 : d\epsilon_3 = \left[\sigma_1 - \frac{1}{2}(\sigma_2 + \sigma_3)\right] : \left[\sigma_2 - \frac{1}{2}(\sigma_1 + \sigma_3)\right]$$
$$: \left[\sigma_3 - \frac{1}{2}(\sigma_1 + \sigma_2)\right] \tag{2-31}$$

Example 2-6

A circle of 0.250-in. diameter was printed on a thin sheet of metal prior to a complex stamping operation. After the stamping is completed, the circle has changed into an ellipse whose major and minor axes are 0.325 and 0.275 in. respectively.

a) Determine the effective strain, $\bar{\epsilon}$, in the region of the ellipse.

b) If we assume that a condition of plane stress prevailed during the operation ($\sigma_z = \tau_{yz} = \tau_{zx} = 0$), where z is the sheet-thickness direction, and the ratio $\alpha = \sigma_2/\sigma_1$ remained constant during deformation, find the ratio of $\sigma_1/\bar{\sigma}$ that must have existed.

Solution

a) The three principal strains are found from

$$\epsilon_1 = \ln\left(\frac{0.325}{0.250}\right) = 0.262$$

$$\epsilon_2 = \ln\left(\frac{0.275}{0.250}\right) = 0.095$$

Since $\epsilon_1 + \epsilon_2 + \epsilon_3 = 0$ from volume constancy, $\epsilon_3 = -0.357$. Using Eq. (2-21),

$$\bar{\epsilon} = \left[\frac{2}{3}(\epsilon_1^2 + \epsilon_2^2 + \epsilon_3^2)\right]^{1/2} = \left[\frac{2}{3}(0.205)\right]^{1/2} \cong 0.370$$

Compare these findings with Eq. (2-23).

†See R. Hill, *Plasticity* (Oxford: Clarendon Press, 1950), p. 50.

b) From the flow rules, using Eq. (2-29), with $\sigma_3 = 0$,

$$\epsilon_1 = \frac{d\bar{\epsilon}}{\bar{\sigma}}\left[\sigma_1 - \frac{1}{2}\sigma_2\right]$$

$$\epsilon_2 = \frac{d\bar{\epsilon}}{\bar{\sigma}}\left[\sigma_2 - \frac{1}{2}\sigma_1\right]$$

or

$$\frac{\epsilon_1}{\epsilon_2} = \frac{0.262}{0.095} = \frac{2\sigma_1 - \sigma_2}{2\sigma_2 - \sigma_1} = 2.76$$

so

$$5.52\sigma_2 - 2.76\sigma_1 = 2\sigma_1 - \sigma_2$$

$$\sigma_2 = \frac{4.76}{6.52}\sigma_1 = 0.73\sigma_1$$

Now, using Eq. (2-16),

$$\bar{\sigma} = \frac{1}{\sqrt{2}}[(\sigma_1 - 0.73\sigma_1)^2 + (0.73\sigma_1)^2 + (-\sigma_1)^2]^{1/2}$$

$$\bar{\sigma} = \frac{1}{\sqrt{2}}[0.073\sigma_1^2 + 0.533\sigma_1^2 + \sigma_1^2]^{1/2} = \frac{1}{\sqrt{2}}[1.606\sigma_1^2]^{1/2}$$

$$\bar{\sigma} = 0.896\sigma_1 \quad \text{or} \quad \frac{\sigma_1}{\bar{\sigma}} = 1.116$$

2-9 THE PRINCIPLE OF NORMALITY†

One interpretation of the flow rules is that the vector sum of the plastic strain increments is normal to the yield surface. This is best shown by Fig. 2-7. In this manner, each incremental strain is treated as a vector, where the principal

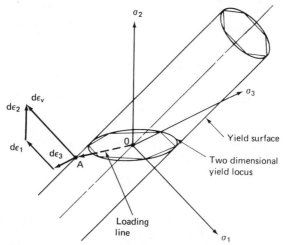

Figure 2-7 Three-dimensional yield surfaces illustrating the coincidence of directions of principal stress and principal strain increments. If the material is loaded to yielding at A, the resulting plastic strain increment may be represented by the vector $d\epsilon_v$ which is normal to the yield surface and is the vector sum of $d\epsilon_1$, $d\epsilon_2$, and $d\epsilon_3$.

†See D.C. Drucker, *Proc. 1st U.S. Nat. Congr. of Appl. Mech.* (1951), p. 487.

coordinates used for stress plots are also used for strain. In Fig. 2-7, $d\epsilon_v$, which is the vector sum of $d\epsilon_1$, $d\epsilon_2$, and $d\epsilon_3$, is normal to the yield surface. With isotropic solids, the principal stress and strain directions coincide. Figure 2-8 shows a portion of a yield locus where the projection of the total strain vector in two-dimensional space is indicated. As a consequence of this, the relation

$$\frac{d\epsilon_1}{d\epsilon_3} = -\frac{\partial\sigma_3}{\partial\sigma_1} \qquad (2\text{-}32)$$

results, where $\partial\sigma_3/\partial\sigma_1$ is the slope of the yield locus at the point of yielding along the loading line. This is very useful in constructing experimental yield loci.†

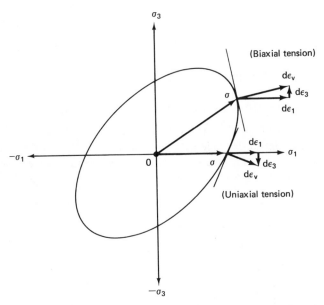

Figure 2-8 Illustration of normality as related to a yield locus. Note that the ratio of $d\epsilon_3$ to $d\epsilon_1$ is given by the normal to the yield locus at the point of yielding.

Example 2-7

A thin sheet is subjected to biaxial tension, $\sigma_1 \neq \sigma_2 \neq 0$ and $\sigma_3 = 0$. The principal strains in the plane of the sheet are $\epsilon_2 = -\frac{1}{4}\epsilon_1$. Using the principle of normality, determine the ratio of σ_2/σ_1 that induced this strain ratio,

a) using the von Mises criterion, and
b) using the Tresca criterion.

†See W. A. Backofen, *Deformation Processing* (Reading, Mass.: Addison-Wesley, 1972), p. 58–72.

Solution

a) Using Eq. (2-29)

$$\frac{\epsilon_2}{\epsilon_1} = \frac{\sigma_2 - \frac{1}{2}\sigma_1}{\sigma_1 - \frac{1}{2}\sigma_2} = -0.25$$

since $\sigma_3 = 0$. Then

$$2\sigma_2 - \sigma_1 = -\frac{1}{4}(2\sigma_1 - \sigma_2)$$

$$2\sigma_2 - \sigma_1 = -\frac{1}{2}\sigma_1 + \frac{1}{4}\sigma_2$$

or

$$\frac{7}{4}\sigma_2 = \frac{1}{2}\sigma_1$$

so

$$\frac{\sigma_2}{\sigma_1} = \frac{2}{7}.$$

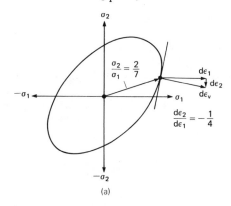

(a)

b) With the Tresca criterion, the yield locus is as follows:

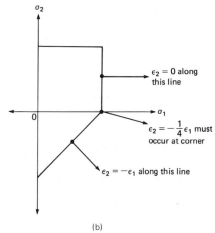

(b)

2-10 MOHR'S CIRCLES FOR STRESS
AND INCREMENTAL PLASTIC STRAIN

The hydrostatic component of the stress state does no plastic work since it causes no expansion of the yield surface; thus, the value of σ_m on a stress circle must coincide with the zero of the circle based upon strain increments. Figure 2-9 shows this result for uniaxial tension. Thus, once the stresses are known, the *ratios* of the plastic strain increments are uniquely defined. The reverse is not true, however, since many stress states that vary only by the magnitude of σ_m would give the same strain ratios.

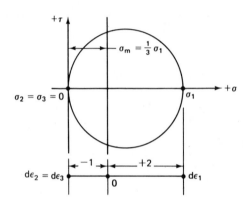

Figure 2-9 Mohr's stress circle for uniaxial tension coincides with Mohr's circle for incremental strains where σ_m is aligned with $d\epsilon = 0$.

APPENDIX

DERIVATION OF THE VON MISES
EFFECTIVE-STRAIN FUNCTION

The effective-strain function should be defined such that the incremental work per unit volume may be expressed as

$$dw = \bar{\sigma}\, d\bar{\epsilon} = \sigma_1\, d\epsilon_1 + \sigma_2\, d\epsilon_2 + \sigma_3\, d\epsilon_3 \qquad \text{(A-1)}$$

For simplicity, consider a state of plane stress where $\sigma_3 = 0$; thus,

$$\bar{\sigma}\, d\bar{\epsilon} = \sigma_1\, d\epsilon_1 + \sigma_2\, d\epsilon_2 = \sigma_1\, d\epsilon_1(1 + \alpha p) \qquad \text{(A-2)}$$

where

$$\alpha = \frac{\sigma_2}{\sigma_1} \quad \text{and} \quad p = \frac{d\epsilon_2}{d\epsilon_1} \qquad \text{(A-3)}$$

Then,

$$d\bar{\epsilon} = d\epsilon_1\left(\frac{\sigma_1}{\bar{\sigma}}\right)(1 + \alpha p) \qquad \text{(A-4)}$$

From the flow rules, with $\sigma_3 = 0$,

$$p = \frac{d\epsilon_2}{d\epsilon_1} = \frac{\sigma_2 - \sigma_1/2}{\sigma_1 - \sigma_2/2} = \frac{2\alpha - 1}{2 - \alpha} \qquad \text{(A-5)}$$

or
$$\alpha = \frac{2p + 1}{2 + p} \tag{A-6}$$

Using Eq. (A-6) in (A-4),

$$d\bar{\epsilon} = d\epsilon_1 \left(\frac{\sigma_1}{\bar{\sigma}}\right)[2(1 + p + p^2)/(2 + p)] \tag{A-7}$$

With $\sigma_3 = 0$, the von Mises expression for $\bar{\sigma}$ is

$$\bar{\sigma} = \frac{1}{\sqrt{2}}[2(\sigma_1^2 + \sigma_2^2 - \sigma_1\sigma_2)]^{1/2} \tag{A-8}$$

and since $\alpha = \sigma_2/\sigma_1$, Eq. (A-8) becomes

$$\bar{\sigma} = \sigma_1(1 - \alpha + \alpha^2)^{1/2} \tag{A-9}$$

Substituting Eq. (A-6) into (A-9) and rearranging gives

$$\frac{\sigma_1}{\bar{\sigma}} = \frac{2 + p}{\sqrt{3}} \left(\frac{1}{1 + p + p^2}\right)^{1/2} \tag{A-10}$$

Introducing Eq. (A-10) into (A-7) gives

$$d\bar{\epsilon} = \frac{2d\epsilon_1}{\sqrt{3}}(1 + p + p^2)^{1/2} \tag{A-11}$$

and since $p = d\epsilon_2/d\epsilon_1$,

$$d\bar{\epsilon} = \frac{2}{\sqrt{3}}(d\epsilon_1^2 + d\epsilon_1 d\epsilon_2 + d\epsilon_2^2)^{1/2} \tag{A-12}$$

Now,

$$d\epsilon_1^2 + d\epsilon_2^2 + d\epsilon_3^2 = d\epsilon_1^2 + d\epsilon_2^2 + (-d\epsilon_1 - d\epsilon_2)^2 = 2(d\epsilon_1^2 + d\epsilon_1 d\epsilon_2 + d\epsilon_2^2) \tag{A-13}$$

and if Eq. (A-13) is introduced into (A-12),

$$d\bar{\epsilon} = \frac{2}{\sqrt{3}} \left[\frac{d\epsilon_1^2 + d\epsilon_2^2 + d\epsilon_3^2}{2}\right]^{1/2} = \left[\frac{2}{3}(d\epsilon_1^2 + d\epsilon_2^2 + d\epsilon_3^2)\right]^{1/2} \tag{A-14}$$

This derivation will also hold for a stress state $\sigma_1', \sigma_2', \sigma_3'$ where $\sigma_3' \neq 0$, since this would be equivalent to a stress state of $\sigma_1 = \sigma_1' - \sigma_3', \sigma_2 = \sigma_2' - \sigma_3', \sigma_3 = \sigma_3' - \sigma_3' = 0$ where $\alpha = \sigma_2/\sigma_1$ and $p = d\epsilon_2/d\epsilon_1 = (\sigma_2 - 1/2\sigma_1)/(\sigma_1 - 1/2\sigma_2)$ as used earlier.

The equivalence of Eqs. 2-19 and 2-20 can be proved by expanding Eq. 2-19.

$$d\bar{\epsilon} = \frac{\sqrt{2}}{3}[(d\epsilon_2 - d\epsilon_3)^2 + (d\epsilon_3 - d\epsilon_1)^2 + (d\epsilon_1 - d\epsilon_2)^2]^{1/2}$$

$$= \frac{\sqrt{2}}{3}[2(d\epsilon_1^2 + d\epsilon_2^2 + d\epsilon_3^2) - 2(d\epsilon_2 d\epsilon_3 + d\epsilon_3 d\epsilon_1 + d\epsilon_1 d\epsilon_2)]^{1/2} \tag{A-15}$$

Expressing the constancy of volume as $d\epsilon_1 = -d\epsilon_2 - d\epsilon_3, d\epsilon_2 = -d\epsilon_3 - d\epsilon_1$, and $d\epsilon_3 = -d\epsilon_1 - d\epsilon_2$ and substituting,

$$d\epsilon_1^2 + d\epsilon_2^2 + d\epsilon_3^2 = (-d\epsilon_2 - d\epsilon_3)^2 + (-d\epsilon_3 - d\epsilon_1)^2 + (-d\epsilon_1 - d\epsilon_2)^2$$

or

$$d\epsilon_1^2 + d\epsilon_2^2 + d\epsilon_3^2 = 2(d\epsilon_1^2 + d\epsilon_2^2 + d\epsilon_3^2) + 2(d\epsilon_2 d\epsilon_3 + d\epsilon_3 d\epsilon_1 + d\epsilon_1 d\epsilon_2) \tag{A-16}$$

Rearranging,

$$2(d\epsilon_2\, d\epsilon_3 + d\epsilon_3\, d\epsilon_1 + d\epsilon_1\, d\epsilon_2) = -(d\epsilon_1^2 + d\epsilon_2^2 + d\epsilon_3^2) \tag{A-17}$$

Finally, substitution of (A-17) into (A-15) gives

$$d\bar{\epsilon} = \left[\frac{2}{3}(d\epsilon_1^2 + d\epsilon_2^2 + d\epsilon_3^2)\right]^{1/2}$$

which is identical to Eq. 2-20.

PROBLEMS

2-1. The area of each face of a metal cube is 2 in.^{2} and this metal has a yield stress in pure shear, k, equal to 20,000 psi. Lubrication is such that normal loading on a face may be assumed frictionless.

(a) If tensile loads of 10,000 and 20,000 pounds are applied in two directions (say, x and y), what is the magnitude of the tensile *load* that must be applied in the z direction to cause yielding according to the Tresca criterion?

(b) If the 10,000- and 20,000-pound loads were compressive, what tensile load in the z direction would cause yielding according to the Tresca criterion?

2-2. A thin-walled tube, with closed ends, is made of a metal whose tensile yield strength is 40,000 psi. The tube has an outer diameter of 3 in. and a wall thickness of .025 in. After applying an *axial compressive load* of 500 pounds to the ends, the tube is pressurized internally. What pressure, P, would cause yielding according to: (a) Tresca?   (b) von Mises?

2-3. A pressure vessel, in the form of a cylinder with hemispherical ends has a radius of 2 ft and is to be made from a metal whose $k = 80$ ksi. The maximum internal pressure intended during use is 5 ksi. If no section of the vessel is to yield, what *minimum* wall thickness should be specified according to: (a) the Tresca criterion? (b) the von Mises criterion?

2-4. Combined tension-torsion stressing is applied to a thin-walled tube. Using the magnitudes of principal stresses in terms of the axial (applied) tensile stress, σ, and the shear stress, τ (applied perpendicular to the axis), determine a relationship between σ, τ, and Y (tensile yield stress) for yielding using (a) the Tresca, and (b) the von Mises yield criteria.

2-5. For a plane-strain compression test, with compressive load F, strip thickness h, width w, and width of indenter b, (see Fig. 3-10) and assuming friction is negligible and the dimension w remains constant during deformation, use the von Mises criterion to determine:

(a) $\bar{\epsilon}$ as a function of ϵ_y.

(b) $\bar{\sigma}$ as a function of σ_y.

(c) An expression of work per unit volume in terms of ϵ_y and σ_y.

(d) An appropriate expression in the form $\sigma_y = f(K, \epsilon_y, n)$ assuming $\bar{\sigma} = K\bar{\epsilon}^n$.
    Note: y is the loading direction!

2-6. A yield criterion is proposed as follows: "Yielding will occur when the *sum* of the two largest shear stresses reaches a critical value," i.e., when

$$(\sigma_1 - \sigma_3) + (\sigma_1 - \sigma_2) = C \text{ if } (\sigma_1 - \sigma_2) \geq$$
$$(\sigma_2 - \sigma_3)$$

or when $(\sigma_1 - \sigma_3) + (\sigma_2 - \sigma_3) = C \text{ if } (\sigma_1 - \sigma_2) \leq$
$$(\sigma_2 - \sigma_3)$$

where $\sigma_1 > \sigma_2 > \sigma_3$ and $C = 2Y$, $Y = $ tensile yield stress.

(a) Is this criterion satisfactory for an isotropic solid where Y is insensitive to the magnitude of σ_m? Justify your answer.

(b) For the case of plane stress (say $\sigma_z = 0$), draw the yield locus in σ_x-σ_y space for this criterion. Sketch the locus for the Tresca criterion on this same plot.

(c) Where $\sigma_z = 0$, what are the values of σ_x and σ_y at yielding if this material is subjected to:
   i) Plane strain where $\epsilon_y = 0$ and $\epsilon_x > 0$?
   ii) axisymmetric flow where $\epsilon_y = \epsilon_z = -\frac{1}{2}\epsilon_x$ and $\epsilon_x > 0$?

2-7. Consider the stress states,

$$\begin{vmatrix} 15 & 3 & 0 \\ 3 & 10 & 0 \\ 0 & 0 & 5 \end{vmatrix} \text{ and } \begin{vmatrix} 10 & 3 & 0 \\ 3 & 5 & 0 \\ 0 & 0 & 0 \end{vmatrix}$$

(a) Find σ_m for each condition.

(b) Find the deviatoric stresses in the normal directions for each condition.

(c) What is the sum of the deviatoric stresses for each case?

2-8. A thin-walled tube with closed ends is made from a metal whose $Y = 40$ ksi. The tube is 8 ft long, has a wall thickness of 0.050 in., and a mean diameter of 3 in. In service, the tube will experience a maximum axial tensile load of 2000 lbf, a maximum torque of 2000 lb-in., and will also be pressurized internally. What minimum internal pressure will cause yielding according to: (a) the von Mises criterion?   (b) the Tresca criterion?

2-9. Consider the cases of a) pure shear, b) uniaxial tension, and c) a triaxial condition where $\sigma_1 > \sigma_2 > \sigma_3$. For each case compare the ratio of $\bar{\sigma}$, the effective stress, and τ_{max}.

2-10. A cube of metal, having a yield strength Y of 345 MPa, is subjected to two perpendicular normal tensile stresses, σ_1 and $\sigma_3 = -\sigma_1/2$.
(a) Find the ratio of $d\epsilon_1/d\epsilon_2$.
(b) What is τ_{max} at the onset of yielding using the von Mises criterion?
(c) Repeat (b) where $\sigma_3 = +\sigma_1/2$.

2-11. Three principal stresses (in ksi) are applied to a solid where $\sigma_1 = 60$, $\sigma_2 = 30$ and $\sigma_3 = 0$.
(a) What is the ratio $d\epsilon_1/d\epsilon_3$?
(b) If a fluid pressure produces an all-around hydrostatic stress of -40 ksi that is superimposed upon the original stress state, how does the ratio in (a) change? Explain this result.

2-12. A cube of metal having a *constant* yield stress Y of 300 MN/m² (such behavior is called *rigid-plastic* since Y would not change after initial yielding) experiences a stress state of σ_1, $\sigma_2 = 0.3\sigma_1$ and $\sigma_3 = -0.5\sigma_1$. If the stresses are gradually increased in these constant ratios, find σ_1 at yielding: (a) using the von Mises criterion, and (b) using the Tresca criterion.

2-13. For plane-strain deformation ($d\epsilon_2$ and $\sigma_3 = 0$), sketch the von Mises yield locus for the first quadrant in σ_1-σ_2 stress space.
   (a) Indicate how $d\epsilon_1$ and $d\epsilon_2$ would appear according to normality.
   (b) Determine the plastic work per unit volume as a function of principal stresses and strain increments.
   (c) Explain your answer in (b) with reference to the plot in (a).

2-14. A long, thin-walled tube with capped ends is pressurized internally until it yields. As pressure is increased, explain what happens to the length of the tube because of plastic effects only.

2-15. A proposed criterion is that yielding will occur when the diameter of the largest Mohr's circle plus half the diameter of the second largest Mohr's circle reaches a critical value.
   (a) Plot the yield locus in σ_x-σ_y space ($\sigma_z = 0$), where σ_x, σ_y, and σ_z are principal stresses.
   (b) In terms of the uniaxial tensile yield stress, Y, what stress state will cause yielding for the case where $\epsilon_y = 0, \epsilon_x > 0$, and $\sigma_z = 0$?

Work hardening

3-1 INTRODUCTION

When metals are deformed plastically at temperatures below those at which recrystallization occurs, they are said to be *cold worked.* Besides the shape change that results, strength and hardness usually increase and the terms *work hardening* or *strain hardening* refer to such strengthening by cold working. Another consequence is the lessening of the remaining ductility.

The most common method used to describe the work-hardening behavior of ductile metals in a quantitative way is by means of a tensile test. Other tests that have been used are direct compression, torsion, balanced biaxial or bulge testing, and plane-strain compression. Because it is the simplest, the tensile test has found major use. For completeness, the various mechanical properties determined from such a test will be defined, but the major emphasis in this chapter will be placed upon the change in yield strength that results from cold working. It is this aspect of work hardening that is of greatest interest in plasticity studies.

3-2 UNIAXIAL TENSILE TEST

The results determined by tensile testing are influenced by the conditions that prevail during such a test; in general, the following are typical:

1. Strain rate is of the order of 10^{-2} to $10^{-4} \sec^{-1}$.
2. Temperature is between 20 and 30°C.

3. Measurements are restricted to a gage section that experiences a state of uniaxial tensile stress during uniform deformation.

The effects of strain rate and temperature are discussed further in Chap. 5.

The basic information obtained is the load or force, F, required to cause a given extension, $\Delta\ell$. From these data, values of stress and strain are computed since they provide more general information. Figure 3-1a is a schematic of a typical load-extension curve. Since the elastic behavior occurs over a very small range of extension as compared with the full test, it appears as a nearly vertical line if all data are plotted to a common scale. To display such behavior, Fig. 3-1b shows the elastic region using an expanded scale along the abscissa.

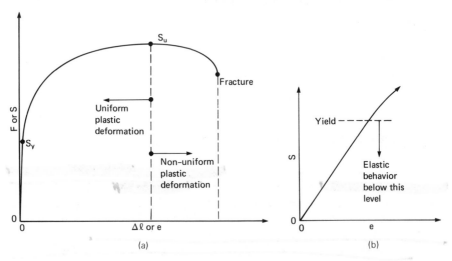

Figure 3-1   (a) Load-extension or nominal stress-strain plot for a ductile metal during a tension test. (b) The expansion of the strain axis to illustrate the elastic region.

3-3 MECHANICAL PROPERTIES

Before determining the traditional properties indicated by a tensile test, the data from Fig. 3-1 are converted to a plot of nominal or engineering stress and strain. These are defined as

$$S = \frac{F}{A_o} \tag{3-1}$$

and
$$e = \frac{\Delta\ell}{\ell_o} = \frac{\ell - \ell_o}{\ell_o} \tag{3-2}$$

where F and $\Delta\ell$ are any pair of coordinate points on Fig. 3-1 and A_o and ℓ_o are the original area and gage length of the unloaded specimen. Since the conversion from load-extension to nominal stress-strain utilizes the constants A_o and ℓ_o, this amounts to nothing more than a scale factor adjustment. Consequently, the shape of the $S\text{-}e$ curve is identical to the $F\text{-}\Delta\ell$ curve. For that reason, Fig. 3-1 carries a dual notation on the axes. When a particular level of stress is reached, plastic flow begins; this stress is called the yield strength and is defined as

$$Y = \frac{F_y}{A_o} = S_y \tag{3-3}$$

the subscript indicating a particular point on the curve.

Tensile strength, also called *ultimate strength* is defined as,

$$S_u = \frac{F_u}{A_o} \tag{3-4}$$

where F_u is the maximum or ultimate load carried by the specimen.

Ductility, which is a measure of the extent to which a metal can be deformed plastically, is commonly defined by two parameters; both are based upon measurements made *after* the specimen has fractured. They are

$$\% \text{ Elongation} = \frac{100(\ell_f - \ell_o)}{\ell_o} \equiv E\ell \tag{3-5}$$

and
$$\% \text{ Reduction of area} = \frac{100(A_o - A_f)}{A_o} \equiv A_r \tag{3-6}$$

where the subscript f relates to the values at fracture. Although standard values of ℓ_o and A_o are generally used, it should be realized that both percent elongation ($E\ell$) and area reduction (A_r) will vary if non-standard values of ℓ_o and A_o are used. For this reason, these properties are not really fixed for a given metal. A_r, the reduction of area at fracture in a tension test, should not be confused with the area reduction in processing. In this text, such deformation is denoted by the symbol r and is defined by

$$r = \frac{A_o - A}{A_o} \tag{3-7}$$

where any given percent cold working is simply r times 100.

The yielding behavior of an annealed brass and a low-carbon steel is illustrated in Fig. 3-2. Brass is typical of most ductile metals in that yielding occurs so gradually that it is difficult to define where the initial plastic deformation takes place. For that reason, the yield stress is usually defined by the offset method. Using a 0.2% offset, a line is constructed parallel to the linear elastic portion of the stress-strain curve but displaced by a strain of 0.002 from the origin, as shown in Fig. 3-2. The 0.2% yield strength, shown as Y_b, is defined by the intersection of the offset line with the stress-strain curve. With the low-carbon steel there is an upper yield point, A, and a lower yield point, B. The latter is much less affected by specimen alignment and rate of loading than is A; for that reason, the stress level at B, shown as Y_s, is used to define the yield strength and there is no need to resort to any offset.

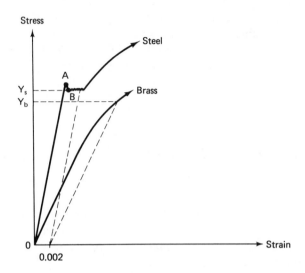

Figure 3-2 Offset method used to define yield strength.

3-4 SOME GENERAL OBSERVATIONS

Returning to Fig. 3-1 it can be seen that after initial yielding has taken place, further plastic deformation requires an increasing load but at a decreasing rate. Work hardening strengthens the material† but at the same time the area is decreasing, and the combined effect of these two factors produces the resulting shape of the load-extension curve. From initial yielding up to ultimate load, the strengthening effect offsets the area reduction; at ultimate load a condition of tensile instability occurs. Note that the deformation of the gage section is uniform up to this point.

At some location along the gage section a local constriction begins and this is called *necking*. From that time on, the continual work hardening in the necked region no longer compensates for the continual reduction of the smallest cross section in the neck. As a consequence, the load required for further extension decreases. It is important to understand that it is the *load*-carrying capacity of this now non-uniform specimen that decreases. Even though the material in the neck has been work hardened more than any other section (and is therefore stronger), the smaller area of the neck causes it to have the lowest load-carrying capacity. With continuing extension, practically all further plastic deformation is restricted to the region of the neck, so a highly non-uniform deformation continues along the full gage section until fracture occurs. The fracture area is then based upon the smallest diameter of the neck.

†Dislocation theory provides a means for explaining this; see reference [1].

3-5 COMPARISON OF NOMINAL
AND TRUE STRESS-STRAIN CURVES

Figure 3-3 shows a nominal stress-strain curve and its counterpart based upon calculations of true stress and true strain using the following definitions:

$$\text{True strain, } d\epsilon = \frac{d\ell}{\ell} \quad \text{or} \quad \epsilon = \ln(1 + e) \tag{3-8}$$

$$\text{True stress, } \sigma = \frac{F}{A} \quad \text{or} \quad \sigma = S(1 + e) \tag{3-9}$$

Here, A is the instantaneous area associated with a particular load.

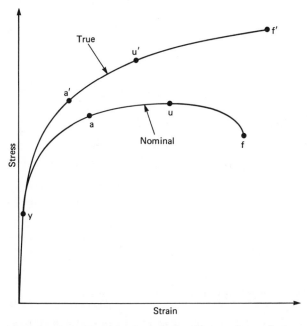

Figure 3-3 Comparison of nominal and true stress-strain curves.

Equation (3-8) is useful only up to the ultimate load. After the onset of necking, the length changes are localized in the neck, so the nominal strain, e, which involves a measurement using the entire gage section, cannot be used to calculate the true strain, ϵ. An alternative expression, indicated earlier by Eq. (1-19), is still useful. This measurement is based upon the minimum cross-sectional area of the neck, and since $d\ell/\ell = -dA/A$,

$$d\epsilon = -\frac{dA}{A} \quad \text{or} \quad \epsilon = \ln\frac{A_o}{A} \tag{3-9a}$$

After necking, the expression $\sigma = F/A$, but *not* $\sigma = S(1 + e)$, gives the *average* true stress in the neck for the direction of loading. This is no longer the effective stress, $\bar{\sigma}$, since the stress state in the neck is now triaxial. A further discussion on this point is given in Sec. 3-7.

When tensile data are used in plasticity studies, true stress and strain are more meaningful than engineering stress and strain; for that reason they will find major use in the rest of this text.

Example 3-1

A tensile specimen of 0.505-in. diameter and 2-in. gage length is subjected to a load of 10,000 lbf. At that instant, the gage length is 2.519 in. Assuming the deformation is uniform to this point,
a) Compute the true stress and true strain.
b) Determine the diameter.

Solution
a) With $d_o = 0.505$ in., $A_o = 0.2$ in.^{2}, and using Eqs. (3-1) and (3-2),

$$S = \frac{10,000}{0.2} = 50,000 \text{ psi}$$

$$e = \frac{2.519 - 2}{2} = 0.259$$

From Eqs. (3-8) and (3-9),

$$\epsilon = \ln(1 + 0.259) = 0.230$$

$$\sigma = 50,000(1.259) = 62,950 \text{ psi}$$

b) $\epsilon = \ln(\ell/\ell_o) = \ln(A_o/A) = 2\ln(d_o/d)$ are all equivalent with constancy of volume. Thus, $\epsilon = 0.230 = 2\ln(0.505/d)$ or $e^{0.115} = 0.505/d$, where e is the base of natural logarithms in this computation. So,

$$d = \frac{0.505}{1.122} = 0.450 \text{ in.}$$

3-6 DETERMINING A WORK-HARDENING EXPRESSION

The curve $oya'u'f'$ in Fig. 3-3 portrays the strain-hardening behavior or the true stress-strain results when plotted on rectangular coordinates. With many ductile metals that have not been cold worked prior to the tensile test (i.e., that are fully annealed), the behavior from initial yield to ultimate load is adequately described by an expression of the form

$$\sigma = K\epsilon^n \tag{3-10}†$$

where, for an induced strain ϵ, the corresponding value of σ is the new yield strength caused by the degree of cold working that induced the strain. Using Eq. (3-7), it can be shown that

$$\epsilon = \ln\left(\frac{1}{1 - r}\right) \tag{3-11}$$

†Note that ϵ is the plastic portion of the total strain both here and in Eq. (3-13).

The important physical consequence of these observations can now be explained. If a certain amount of cold work is induced in a metal, this corresponds to a particular value of r, and with Eq. (3-11) the equivalent value of ϵ is determined. Introducing this value into Eq. (3-10), and assuming that K and n are known, permits a calculation of σ, which is the new yield strength due to the effect of cold working. Via such a procedure, it is possible to quantify reasonably the yield strength as a function of cold working. Note, too, that the conditions which are described by Eq. (3-10) were $\sigma_1 \neq 0, \sigma_2 = \sigma_3 = 0$ and $d\epsilon_1 = -2d\epsilon_2 = -2d\epsilon_3$ where the loading direction is taken as 1 and isotropy and volume constancy are assumed. Using Eqs (2-16), and (2-19) or (2-20) it can be shown that

$$\sigma_1 = \bar{\sigma} \quad \text{and} \quad d\epsilon_1 = d\bar{\epsilon} \qquad (3\text{-}12)$$

Thus, the tensile results are, in fact, descriptive of an effective stress-strain plot. For that reason we can now write Eq. (3-10) as

$$\bar{\sigma} = K\bar{\epsilon}^n = Y \qquad (3\text{-}13)$$

From this point on, this is the expression that we will use as an effective stress-effective strain function.

Example 3-2

The plastic behavior of a certain metal is expressed as $\bar{\sigma} = 100,000\bar{\epsilon}^{0.50}$. If a bar of this metal is uniformly cold worked to an area reduction of $r = 0.3$, estimate the yield strength of the cold-worked piece.

Solution.   With Eq. (3-11), the induced true strain is

$$\epsilon = \ln\left(\frac{1}{1 - 0.3}\right) = 0.357$$

From Eq. (3-10) and with the values of K and n given, the new yield strength, Y, is

$$Y = \sigma = 100,000(0.357)^{0.5} = 59,749 \text{ psi}$$

It is of course essential to determine numerical values of K and n before Eq. (3-13) finds application. If Eq. (3-10) is descriptive of the plastic behavior of the metal, the simplest approach is to plot the σ-ϵ data on logarithmic coordinates, since a power-law expression plots as a straight line on those scales. Figure 3-4 shows experimental values obtained for a commercially pure aluminum specimen which was fully annealed before being subjected to a tensile test. With logarithmic coordinates, there is no zero-zero starting point, so the elastic region, depicted as Zone I, must start at some finite value. As an aside, if the metal follows Hooke's law, the stress-strain relation in this zone obeys $\sigma = E\epsilon$ and the plotted points must form a 45°-line to either axis. Thus, although E is defined by the slope of the line when a plot on rectangular coordinates is constructed, the *slope* of this line on logarithmic coordinates bears no relation to the elastic modulus. Differences in moduli are shown by the relative position of such lines with respect to each other, and their extrapolation to the intersection with unit strain defines the value of E; this is shown on Fig. 3-4.

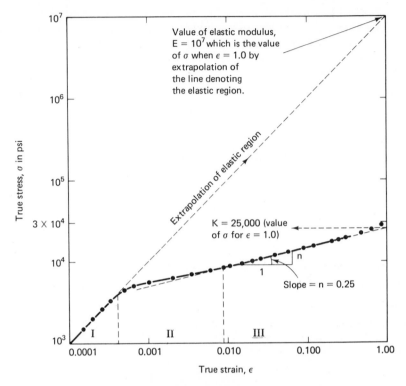

Figure 3-4 True stress-strain behavior of 1100–0 alumimum tested in uniaxial tension when plotted on logarithmic coordinates. Taken from R. M. Caddell and R. Sowerby, "Strain Hardening: An Introduction to a Fundamental Experiment," *Bull. Mech. Engrg. Educ.*, 8 (1969), pp. 31–43.

Zone II represents a transition from elastic to fully plastic behavior and is typical of most ductile metals. Materials such as low-carbon steel, having a pronounced yield point, would exhibit the behavior shown in Fig. 3-5.

Once Zone III is reached, the metal may be viewed as being in a "fully plastic" condition and from this point up to necking the measured test points fall on a straight line. The slope of this line defines the *strain-hardening exponent*, n, and the intersection of this line with unit strain gives the stress value that defines the magnitude of K. This is often called the *strength coefficient*. Thus, we have

$$\sigma = E\epsilon^{1.0} \quad \text{(elastic-Zone I)} \tag{3-14a}$$

$$\sigma = K\epsilon^{n} \quad \text{(plastic-Zone III).} \tag{3-14b}$$

Since no one, simple equation describes the stress-strain behavior of real metals from the onset of loading to fracture, objections are sometimes raised to the use of Eq. (3-10) as a means of describing strain-hardening behavior. Of course, other forms of strain-hardening relations would be subjected to these

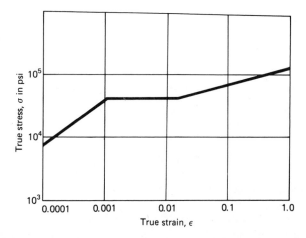

Figure 3-5 True stress-strain behavior of annealed, low-carbon steel tested in uniaxial tension when plotted on logarithmic coordinates.

same objections. It is therefore essential to realize the physical restrictions that must be considered when a power-law form of strain hardening is utilized. These are:

1. Such an equation is quite reliable when the induced strain is greater than 0.04 but less than the strain at which necking begins.
2. Use of this equation to predict the initial yield strength of the metal should be avoided. Instead, a method such as offset should be used.
3. Most metal-working operations impart strains far in excess of 0.04 (this is equivalent to about 4% cold work!), and the exclusion of the elastic and "transition" strain regions leads to little error in this regard.

Since by definition

$$\sigma = \frac{F}{A} = K\epsilon^n \quad \text{and} \quad d\epsilon = \frac{d\ell}{\ell} = -\frac{dA}{A} \tag{3-15}$$

at the instability point (maximum load) where $dF = 0$, it can be shown that $\epsilon_u = n$. Although this is mathematically correct, it is very difficult to measure the true strain at the exact maximum load because of the machine sensitivity that would be required; consequently, a plot such as Fig. 3-4 is the most reliable method for defining n. Often, however, a question can arise concerning the placement of the line that best describes Zone III behavior; the following should assist in this regard.

Example 3-3

Show that at the onset of tensile instability, assuming that plastic behavior is described by $\bar{\sigma} = K\bar{\epsilon}^n$, the true strain at the ultimate load, ϵ_u, equals the strain-hardening exponent, n.

Solution. By definition, $F = \sigma A$, so $dF = \sigma dA + A d\sigma = 0$ at maximum load. So

$$\frac{d\sigma}{\sigma} = -\frac{dA}{A} = d\epsilon \text{ by definition}$$

or

$$\frac{d\sigma}{d\epsilon} = \sigma$$

With $\sigma = K\epsilon^n$,

$$\frac{d\sigma}{d\epsilon} = nK\epsilon^{n-1} = \sigma = K\epsilon^n$$

or

$$\frac{n}{\epsilon} = 1 \quad \text{so} \quad n = \epsilon$$

Since this relates to the condition at ultimate load, $n = \epsilon_u$.

The true stress at ultimate load (note that it is dangerous to call this the "true tensile strength" since "tensile strength" is the maximum engineering stress whereas the maximum true stress occurs at fracture) can be expressed as

$$\sigma_u = K\epsilon_u^n = Kn^n \quad \text{since} \quad \epsilon_u = n \tag{3-16}$$

From Eqs. (3-4), (3-9), and (3-16), at maximum load

$$F_u = S_u A_o = \sigma_u A_u = (Kn^n)A_u \tag{3-17}$$

Therefore

$$S_u = (Kn^n)\frac{A_u}{A_o}$$

From Eq. (3-9a),

$$\frac{A_u}{A_o} = e^{-\epsilon_u} \tag{3-18}$$

so with $\epsilon_u = n$,

$$S_u = K\left(\frac{n}{e}\right)^n \tag{3-19}$$

where e is the *base of natural logarithms* in Eqs. (3-18) and (3-19).

Once a straight line is fitted to test points, thereby defining K and n, S_u can be calculated. If this varies from the *measured* value of tensile strength (say by $\pm 3\%$), it is probable that improper weight has been given to points in the Zone II region; a slight adjustment of the line would be needed. This technique provides a very useful check in assessing the proper values of K and n.

Example 3-4

True stress-true strain data are plotted on logarithmic coordinates as in Fig. 3-4. A straight line, that seems to fit best the plastic zone, produces the strain-hardening equation, $\bar{\sigma} = 50,000 \, \bar{\epsilon}^{0.25}$. During this test the tensile strength was accurately measured to be 28,000 psi. Do K and n appear to be proper values?

Solution. Using Eq. (3-19), the tensile strength is *calculated* as

$$S_u = 50,000\left(\frac{0.25}{e}\right)^{0.25} = 27,535 \text{ psi}$$

Thus the percent variation involved $= (28,000 - 27,535)/28,000 = 1.66\%$; so K and n are certainly reasonable.

It has been implied that the parameters K and n are material constants, and it is essential that this concept be understood thoroughly. One cannot, for example, assume that the magnitude of these parameters is fixed for a metal whose structure can be significantly altered by heat treatment. If a piece of SAE 1020 steel were fully annealed while a second piece had been austenitized and oil quenched, different values of K and n would be found for these two specimens. Again, specimens of 2024 aluminum that were solution treated and age hardened would produce different values of K and n when compared with this same metal in an overaged condition. In effect, each chemical composition and condition of microstructure must be viewed as a *different* metal as far as K and n are concerned. What must be realized, however, is that the value of the parameters should be determined with a metallic specimen that *contains no effect of work hardening prior to the tensile deformation itself.* After all, K and n are the very parameters used to describe the work-hardening characteristics. To account for the effects of cold working that may have been induced initially, Swift proposed an expression equivalent to

$$\bar{\sigma} = K(\bar{\epsilon}_o + \bar{\epsilon}_a)^n \qquad (3\text{-}20)$$

where $\bar{\epsilon}_o$ is the strain due to prior cold working and $\bar{\epsilon}_a$ is the strain due to subsequent plastic deformation. If one realizes that $\bar{\epsilon}$ in Eq. (3-13) is the total plastic strain, then Eqs. (3-13) and (3-20) are really equivalent. Many other empirical expressions have been proposed in this regard; Johnson and Mellor [2] present some of them.

Example 3-5

a) The strain-hardening behavior of an annealed, low-carbon steel is $\bar{\sigma} = 100,000\,\bar{\epsilon}^{0.2}$. If a bar of this metal is initially cold worked 20%, followed by additional cold work of 30%, determine the probable yield strength of the final bar.

b) Suppose another bar of this annealed material is initially cold worked some unknown amount. It is then subjected to 15% additional cold work and its yield strength is measured as 75,000 psi. What was the unknown amount of cold work induced prior to the 15% added cold work?

Solution

a) From step one, $\bar{\epsilon} = \ln(1/1 - 0.2) = 0.223 = \bar{\epsilon}_o$ in Eq. (3-20). From the second step, $\bar{\epsilon} = \ln(1/1 - 0.3) = 0.357 = \bar{\epsilon}_a$ in Eq. (3-20)

The total induced strain $= 0.223 + 0.357 = 0.580$, so, $Y = \bar{\sigma} = 100,000(0.58)^{0.2}$ $= 89,678$ psi.

b) Since Y is known, the *total* induced strain is found first,

$$75,000 = 100,000(\bar{\epsilon})^{0.2}$$

so $\bar{\epsilon} = 0.75^5 = 0.237$

The strain due to the 15% cold work $= \ln(1/1 - 0.15) = 0.163$. From $\bar{\epsilon}_o + \bar{\epsilon}_a$ $= \bar{\epsilon}_{total}$, $\bar{\epsilon}_o = 0.237 - 0.163 = 0.074$ and $0.074 = \ln(1/(1 - r_1))$ where r_1 is the initial cold work. $r_1 = 0.071$ or about 7% cold work.

This illustrates the additive property of true strains and it should be noted that percent cold work is not an additive property. For instance, if the total cold work were calculated from the total induced strain, as $0.237 = \ln(1/1 - r)$, $r = 0.21$ or 21%, the original amount computed from $r_t - r_2 = 21 - 15 = 6\%$. This answer would be incorrect. Although it is close to the correct answer of 7% *in this example*, such a procedure will lead to far greater variation if different numbers are used. The reader can check this if desired.

3-7 BEHAVIOR AFTER NECKING

In Sec. 3-5 it was noted that once necking begins, the true stress, $\sigma = F/A$, is no longer equal to the effective stress, $\bar{\sigma}$; therefore it does not correctly describe subsequent strain hardening. The reason for this is the triaxial stress state that develops in the neck whereas, prior to necking, a uniaxial stress existed. As the smallest section of the neck elongates under continued displacement, its lateral contraction is restrained by adjacent sections that are more lightly stressed (i.e., these areas are a little larger). This constraint causes radial and circumferential tension in the neck where these stresses increase from zero at the surface to a maximum in the interior. To maintain flow, the axial stress must increase from $\bar{\sigma}$ at the surface to a maximum at the centerline. Thus, the average axial true stress, $\sigma = F/A$, rises above $\bar{\sigma}$. This is shown on Fig. 3-4 where after necking, the plotted points rise above the line described by Eq. (3-10).

Bridgman [3] analyzed this problem and developed an expression for $\bar{\sigma}/\sigma$ as a function of a/R where $2a$ is the minimum diameter in the neck and R is the radius of curvature; Fig. 3-6a shows a plot of this function while Fig. 3-6b illustrates the neck *geometry* and the axial stress distribution. When R is infinite, no correction is necessary (i.e., $\bar{\sigma} = \sigma$), but as a neck develops and becomes sharper, R decreases and $\bar{\sigma}/\sigma$ drops below unity. Use of the Bridgman correction together with accurate measurements of a and R permit determining

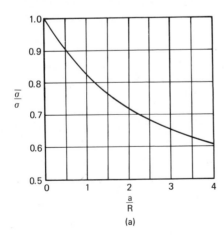

(a)

Figure 3-6 (a) Relation of effective stress, $\bar{\sigma}$, and the average axial stress, σ, in a necked tensile bar after Bridgman [3].

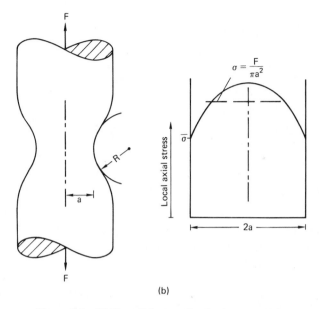

Figure 3-6 (b) The axial stress distribution in the neck.

values of $\bar{\sigma}$ and $\bar{\epsilon}$ at strains in excess of ϵ_u or n. However, if the neck becomes very severe, the hydrostatic tension at the centerline will cause voids to form. This then decreases the true load-carrying cross section below the value of πa^2 inferred from external measurements. In fact, it is the linkage of such voids that causes eventual fracture of ductile tensile specimens.

3-8 DIRECT COMPRESSION

An apparent solution that avoids necking is to conduct tests in uniaxial compression. Difficulties arise in this situation because homogeneous compression is not as simple as it appears. Frictional effects at the interface between the specimen and the loading surfaces can lead to what is called *barreling*, as indicated on Fig. 3-7.

As load is applied to a right circular specimen of original height and diameter of h_o and d_o, respectively, the height becomes smaller and the area larger. At any particular load, the uniaxial stress and strain would come from

$$\sigma = \frac{F}{A} \quad \text{and} \quad \epsilon = \ln\left(\frac{h}{h_o}\right)$$

both of these quantities being negative in sign. Because of the influence of friction at the workpiece-loading surface interface, the ends of the specimen are restricted from free radial expansion. As a consequence, cone-shaped zones (shaded regions of Fig. 3-7) of relatively undeformed metal occur at each end of

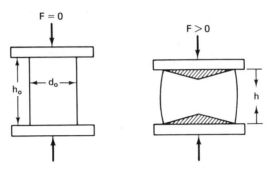

Figure 3-7 Direct compression showing the formation of dead-metal zones and barrelling.

the specimen and, in time, barreling becomes pronounced. The stresses and strains under these conditions are no longer uniform.

If the effects of friction can be minimized, uniformity can be extended to greater deformations. It appears that the method of Cook and Larke [4], later modified by Watts and Ford [5], provides the best means for obtaining true stress-strain compression data; it is, however, more involved than is the tensile test. In general, the method utilizes cylinders of equal diameters but varying heights, such that the d_o/h_o ratio varies from 0.5 to 3.0. The *modified* approach is to start with well-lubricated ends and apply a particular load to a specimen, remove the load, and measure the new height. Upon relubrication, the specimen is subjected to an increased load, unloaded, measured, and so on. This same approach is followed with each specimen where the particular *load* levels are exactly duplicated. Figure 3-8 shows how the results are plotted. For the same load, the actual reduction in height is plotted against the diameter-to-height ratio for each cylinder. A line drawn through the points is extended to intersect a d_o/h_o ratio of zero; this is equivalent to testing a specimen of infinite initial height, where the end effects of friction would be restricted to

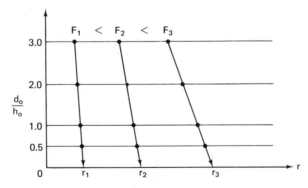

Figure 3-8 Watts and Ford [5] method to correct for end effects in compressive loading.

a small region of the full test height. In this way, it is theorized that the major portion of the specimen would deform uniformly.

With the value of r determined for a zero d_o/h_o ratio, the "uniform" true strain can be calculated using Eq. (3-11). As the original cross-sectional area based upon d_o is known, and assuming the volume remains constant, the theoretical uniform change in area can be computed for a given load. The true stress follows by dividing this load by this computed area; thus, a "uniform" true stress-strain combination is obtained. With a number of such points, the curve in compression can be plotted.

Examples exist in the literature to illustrate that, for many metals, the tensile and compressive values of true stress-strain are practically equivalent if sufficient care is exercised in obtaining the data [6]†. The important conclusion to be drawn is that the strain-hardening equation obtained from uniaxial tension can be employed at strains in excess of the necking tensile strain with some degree of confidence, especially when compressive deformation predominates.

3-9 BULGE TEST OR BALANCED BIAXIAL TENSION

In the bulge test, a thin disc of sheet metal is clamped around its periphery and then subjected to an increasing fluid pressure applied to one side as shown in Fig. 3-9. As the sheet bulges, the region in the vicinity of the dome becomes nearly spherical in shape and the tensile stresses in the plane of the sheet are

$$\sigma_\theta = \sigma_\phi = \frac{P\rho}{2t} \quad \text{by symmetry} \tag{3-21}$$

where P, ρ, and t are the instantaneous pressure, radius of curvature, and thickness, respectively. The tensile strains ϵ_θ and ϵ_ϕ are equal because of symme-

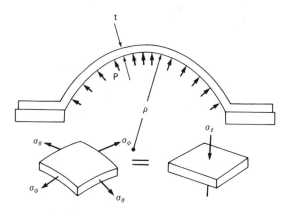

Figure 3-9   Schematic of a bulge test showing equivalent stress states.

†There are however, many exceptions.

try, and because of volume constancy the thickness strain is

$$\epsilon_t = -2\epsilon_\theta = -2\epsilon_\phi = \ln\left(\frac{t}{t_o}\right) \tag{3-22}$$

where t_o is the initial thickness. In practice, ϵ_θ or ϵ_ϕ is computed from pertinent measurements, then ϵ_t is determined from Eq. (3-22).

The actual stress state, $\sigma_\phi = \sigma_\theta$ and $\sigma_z = 0$, is equivalent to $\sigma_\phi' = \sigma_\theta' = 0$, $\sigma_z' = -\sigma_\theta$, so the results of such a test are equivalent to the *compressive* stress-strain behavior in the thickness direction. For isotropic metals, these results are practically identical to the uniaxial tension or direct compression behavior found with specimens made from the same bulk material. An advantage of the bulge test is that it can be carried out to strains far greater than the instability strain in tension (i.e., n), and is not complicated by friction as is the compression test.

3-10 PLANE-STRAIN COMPRESSION TEST

Figure 3-10 is a schematic of the essentials involved where $w/b > 6$ and $2 < b/t < 4$ are recommended. This also uses relatively thin materials. As load is applied, the metal between the indenters is prevented from moving in the w direction by the constraint from the unstressed material adjacent to the indented region. Thus $\epsilon_w = \epsilon_2 = 0$. To obtain the best results, incremental loading is employed where a strain from 2 to 5% is induced per increment. Stress is calculated by dividing a particular load by the area between the indenters, while a corresponding strain is computed from $\epsilon = \ln(t/t_o)$. Because

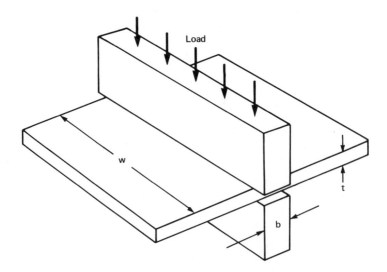

Figure 3-10 Details of a plane-strain compression test.

of plane-strain conditions, the σ-ϵ data as calculated will not duplicate that behavior obtained from uniaxial tension, direct compression, or bulge tests. Using the concept of effective stress and strain, it is often found that a single expression of the form $\bar{\sigma} = K\bar{\epsilon}^n$ results.

Example 3-6

Consider Fig. 3-10 where the loading direction is 1, the width direction 2, and the length direction 3. Express the effective stress, $\bar{\sigma}$, and effective strain, $\bar{\epsilon}$, as functions of the stress and strain measured in the loading direction.

Solution. Because this is plane strain and there is no force applied in direction 3, it must be noted at the outset that $\epsilon_2 = 0$ and $\sigma_3 = 0$, so $\sigma_2 = \frac{1}{2}(\sigma_1 + \sigma_3) = \frac{1}{2}\sigma_1$ and $\epsilon_3 = -\epsilon_1$.

Using Eqs. (2-16) and (2-21),

$$2\bar{\sigma}^2 = \left(\sigma_1 - \frac{\sigma_1}{2}\right)^2 + \left(\frac{\sigma_1}{2}\right)^2 + (\sigma_1)^2 = \frac{6}{4}\sigma_1^2$$

so

$$\bar{\sigma} = \frac{\sqrt{3}}{2}\sigma_1$$

$$\bar{\epsilon} = \left[\frac{2}{3}(\epsilon_1^2 + 0 + \epsilon_3^2)\right]^{1/2} = \left[\frac{2}{3}(2\epsilon_1^2)\right]^{1/2} = \frac{2}{\sqrt{3}}\epsilon_1$$

3-11 TORSION

Although not as widely used as the previous methods, torsion of thin-walled tubes and solid bars has been studied in connection with work-hardening behavior. We include this short qualitative discussion for completeness.

The general approach is to convert torque-angular twist data into shear stress-shear strain data. Then using pertinent assumptions, the equivalence with a σ-ϵ curve follows. Detailed discussions on procedures may be found in references [2], [7], and [8]. Torsional studies do find use in certain hot-working operations where very high strains may be induced.

For determining strain-hardening behavior caused by cold working, the tensile test is the simplest to conduct, and for that reason has found the widest application.

REFERENCES

[1] R. M. Caddell, *Deformation and Fracture of Solids*. Englewood Cliffs, N.J.: Prentice-Hall, 1980, pp. 118–21, 177–78.

[2] W. Johnson and P. B. Mellor, *Engineering Plasticity*. New York: Van Nostrand Reinhold, 1973, pp. 15–17, 118–121.

[3] P. W. Bridgman, *Trans. Amer. Soc. Metals*, 32 (1944), pp. 553–74.

[4] M. Cook and E. C. Larke, *J. Inst. Metals*, 71 (1945), pp. 371–90.

[5] A. B. Watts and H. Ford, *Proc. Inst. Mech. Eng.*, 169 (1955), pp. 1141–49.

[6] E. G. Thomsen, C. T. Yang, and S. Kobayashi, *Mechanics of Plastic Deformation in Metal Processing*. New York: Macmillan, 1965, pp. 106–11.

[7] C. R. Calladine, *Engineering Plasticity*. Elmsford, N.Y.: Pergamon Press, 1969, pp. 52–54.

[8] D. S. Fields and W. A. Backofen, *Proc. ASTM*, 57 (1957), pp. 1259–72.

PROBLEMS

3-1. Show that $\epsilon = \ln(1 + e)$ and $\sigma = S(1 + e)$ as given by Eqs. (3-8) and (3-9); then discuss any practical limitation of these relationships.

3-2. A tensile specimen is machined to a gage diameter of 0.357 in. and is marked with a starting gage length of 2 in. When subjected to a test, the following results were found:

yield load = 2000 lbf
fracture diameter = 0.270 in.
diameter at ultimate load = 0.310 in.
elastic modulus = 25×10^6 psi

After completing this test, you are informed that the tensile specimen had been cold worked some amount *before* it was machined and tested, and that in the annealed state $\bar{\sigma} = K\bar{\epsilon}^n$ with $n = 0.5$.
(a) What is Y for this specimen?
(b) How much strain was induced by the unknown amount of cold work?
(c) What maximum load (i.e., F_u) was reached during the test?

3-3. An annealed brass specimen of 0.505-in. starting diameter supports a maximum tensile load of 120,000 lbf at which point the initial area is reduced 40%. If a second identical specimen were loaded until the induced strain was half the magnitude of n, what load would be needed to reach this condition?

3-4. During a tensile test with a metal that obeys $\bar{\sigma} = K\bar{\epsilon}^n$, the tensile strength is found to be 340 MPa. Reaching the maximum load required an elongation of 30%. From this limited information, compute K and n.

3-5. A thin-walled tube of annealed aluminum is placed over a mandrel that just fits the inner diameter of the tube. The tube is then alternately stretched and compressed, the sequence being a 5% stretch, then compression to the original length. Assuming any induced strain is uniform and $\bar{\sigma} = 25{,}000\bar{\epsilon}^{0.25}$ describes the strain-hardening behavior, how many cycles (take a stretch plus a compression as one cycle) are needed to produce a yield strength of 30 ksi for the tube?

3-6. If $\bar{\sigma} = K\bar{\epsilon}^n$ applies to a metal, show that if it is strained to a level ϵ_1 due to σ_1, the plastic work per unit volume is $\sigma_1\epsilon_1/(1 + n)$.

3-7. A plate of steel is loaded under balanced biaxial tension such that $\sigma_1 = \sigma_2$, $\sigma_3 = 0$. It has a yield stress, Y, of 40,000 psi, Young's modulus is 30×10^6 psi and Poisson's ratio is 0.30. What is the largest fractional volume change, $\Delta V/V$, that could be obtained without yielding? Assume the von Mises criterion prevails.

3-8. For plane-strain compression, as in Fig. 3-10,

(a) Express the work per unit volume, dw, in terms of $\bar{\sigma}$ and $d\bar{\epsilon}$ and compare it with $dw = \sigma_1 d\epsilon_1 + \sigma_2 d\epsilon_2 + \sigma_3 d\epsilon_3$.

(b) If $\bar{\sigma} = K\bar{\epsilon}^n$, express σ_1 (loading direction) as a function of ϵ_1, K, and n.

3-9. The following tensile data were obtained using a steel specimen:

Load (kN)	Min. Diam. (mm)	Neck Radius (mm)	True Strain $\bar{\epsilon}$	True Stress (Apparent) σ-MPa	True Stress (Corrected) $\bar{\sigma}$-MPa
0	8.69	∞	0	0	0
27.0	8.13	∞	0.133	520	520
34.4	7.62	∞			
40.6	6.86	∞			
38.3	5.33	10.3			
29.2	3.81	1.8			

(a) Compute all missing values.

(b) Plot both σ and $\bar{\sigma}$ vs. $\bar{\epsilon}$ on 2 cycle log-log paper and determine the *correct* values of K and n in the expression $\bar{\sigma} = K\bar{\epsilon}^n$. See Fig. 3-6.

(c) How much energy is consumed per unit volume to induce a true strain of 0.35 into the bar?

4

Plastic instability

4-1 INTRODUCTION

The extent to which a metal may be deformed in an acceptable manner can be limited by different criteria. If under compressive loading, the ratio of the height to the cross section of the workpiece is too large, buckling will occur. Such a sidewise deflection prevents uniform compression from occurring. The well-known Euler equation expresses the load to cause buckling as a function of the elastic modulus, the length of the column (i.e., the height), and the smallest value of the second moment of area of the cross section. A similar expression exists if plastic rather than elastic buckling is encountered. At the other extreme or forming limit is the occurrence of fracture where physical separation of the workpiece results. To avoid buckling, a "small" strain event, it is only necessary to ensure that the height to cross section ratio is less than some critical value. Fracture, which results after "large" strains are induced, can be avoided by restricting the degree of strain imparted to the deforming metal. No further comments will be made about these two possible occurrences since the thrust of this chapter has to do with a different phenomenon. Most commonly this is called *plastic instability* since it relates to a condition of plastic deformation at which point the deformation would continue under a falling load or pressure; in essence, the process would become unstable.

4-2 GENERAL APPROACH TO INSTABILITY—UNIAXIAL TENSION

Throughout this chapter it is assumed that the strain-hardening behavior is expressed by $\bar{\sigma} = K\bar{\epsilon}^n$; if other expressions provide a better description they should be used but the general techniques that follow would still be applicable. Often a solution is found for which the effective strain at instability is some function of n. This is useful if one is concerned with the general extent of uniform strain that can be induced before the onset of instability. Again, greater interest may center upon the actual load or pressure at which instability can be expected. As an introduction, consider the case of uniaxial tension with a bar of uniform cross section.

In a tensile test of a ductile metal, the onset of necking or non-uniform deformation coincides with the maximum load or applied force attained during the test; this is the instability condition for such a situation. In terms of true stress and true strain,

$$\bar{\sigma} = \frac{F}{A} \quad \text{or} \quad F = \bar{\sigma}A \tag{4-1}$$

When this maximum load is reached, $dF = 0$, so

$$dF = \bar{\sigma}dA + Ad\bar{\sigma} = 0 \tag{4-2}$$

Rearranging Eq. (4-2) gives

$$\frac{d\bar{\sigma}}{\bar{\sigma}} = -\frac{dA}{A} = d\bar{\epsilon} \tag{4-3}$$

so

$$\frac{d\bar{\sigma}}{d\bar{\epsilon}} = \bar{\sigma} \tag{4-4}$$

With $\bar{\sigma} = K\bar{\epsilon}^n$, Eq. (4-4) becomes

$$\frac{d\bar{\sigma}}{d\bar{\epsilon}} = nK\bar{\epsilon}^{n-1} = K\bar{\epsilon}^n = \bar{\sigma} \tag{4-5}$$

or

$$\bar{\epsilon} = n \tag{4-6}$$

Thus the maximum load, tensile strength, and the onset of necking are all displayed when the effective strain is equal to the strain hardening exponent.

At lower strains ($\bar{\epsilon} < n$) uniform deformation occurs along the full gage section, since the applied load is dictated by the weakest section; if all such sections are equivalent, uniform extension results. If, by chance, any region deformed more than the rest of the gage section, this increased cold work would cause the *load*-carrying capacity of that region to exceed regions elsewhere. As a consequence, further deformation in that more highly deformed region would cease until the remainder of the bar deformed enough to require further increase in the applied load. However, once the effective strain exceeds n and necking begins, the necked region exhibits the lowest *load*-carrying capacity of the entire bar even though it has the highest flow stress since it has experienced the greatest amount of work hardening. With continued loading beyond this point, a decreasing load is needed to cause further deformation of the neck, and deformation outside of this region then ceases. The end result is that uniform elongation is governed by n.

A construction attributed to Considére is often presented to illustrate a graphical approach to instability. In terms of $\bar{\sigma}$ and $\bar{\epsilon}$, this is shown in Fig. 4-1, where the instability point occurs when the magnitude of the slope of the stress-strain curve ($d\bar{\sigma}/d\bar{\epsilon}$) is equal to $\bar{\sigma}$; this gives a length of unity along the strain axis.

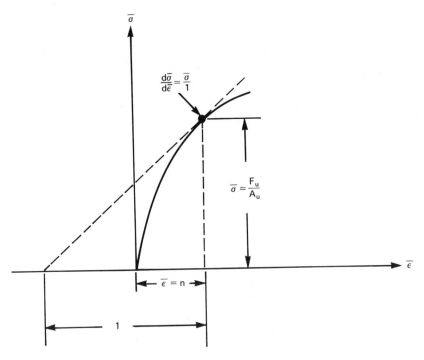

Figure 4-1 Considére construction for instability in uniaxial tension.

4-3 BALANCED BIAXIAL TENSION

This approximates the stretching of a thin sheet. Figure 4-2 illustrates this situation where the loads F_1 and F_2 are equal, F_3 being zero. Each starting area, A_o, upon which the loads will act is simply $t_o w_o$. Because of symmetry, either F_1 or F_2 may be used in the analysis. Instability results when $dF = 0$. Now

$$F_1 = \sigma_1 A_1 \qquad\qquad (4\text{-}7)$$

so

$$dF_1 = 0 = \sigma_1 dA_1 + A_1 d\sigma_1$$

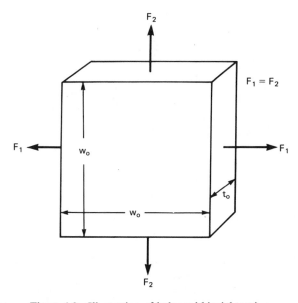

Figure 4-2 Illustration of balanced biaxial tension.

or
$$\frac{d\sigma_1}{\sigma_1} = -\frac{dA_1}{A_1} = d\epsilon_1 \tag{4-8}$$

The von Mises expression for the effective stress with $\sigma_1 = \sigma_2$ and $\sigma_3 = 0$ shows that

$$\bar{\sigma} = \sigma_1 = \sigma_2 \tag{4-9}$$

From the flow rules and Eq. (2-21),

$$\bar{\epsilon} = 2\epsilon_1 = 2\epsilon_2 = -\epsilon_3 \tag{4-10}$$

Substituting Eqs. (4-9) and (4-10) into (4-8) gives

$$\frac{d\bar{\sigma}}{d\bar{\epsilon}} = \frac{\bar{\sigma}}{2} \tag{4-11}$$

and with $\bar{\sigma} = K\bar{\epsilon}^n$, Eq. (4-11) gives

$$\bar{\epsilon} = 2n \tag{4-12}$$

at instability. Note that this is twice the instability strain predicted for uniaxial tension.

4-4 THIN-WALLED SPHERE UNDER INTERNAL PRESSURE

Consider Fig. 4-3, which shows a free body diagram of one half a spherical pressure vessel. Due to symmetry, $\sigma_1 = \sigma_2$ while $\sigma_3 = 0$; here the 1, 2, 3 system corresponds to x, y, and z as shown. From a force balance in the x or 1 direction, $\sigma_1 = Pr/2t$ where P is the internal pressure and r and t the instan-

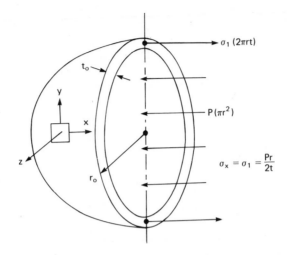

Figure 4-3 Free body diagram of a spherical shell subjected to internal pressure.

taneous radius and wall thickness at pressure P. In this instance, instability occurs at some pressure, so the governing condition will be found when $dP = 0$. The solution is started by noting that

$$P = \frac{2\sigma_1 t}{r} \tag{4-13}$$

Instability occurs when $dP = 0$, so using Eq. (4-13)

$$dP = 0 = \frac{2t d\sigma_1}{r} + \frac{2\sigma_1 dt}{r} - \frac{2t\sigma_1 dr}{r^2} \tag{4-14}$$

which can be written as

$$\frac{d\sigma_1}{\sigma_1} = \frac{dr}{r} - \frac{dt}{t} \tag{4-15}$$

Recognizing that $dr/r = d\epsilon_1$ and $dt/t = d\epsilon_3$ and from symmetry $d\epsilon_1 = d\epsilon_2 = -d\epsilon_3/2$, Eq. (4-15) becomes

$$\frac{d\sigma_1}{\sigma_1} = -\frac{3}{2} d\epsilon_3 \tag{4-16}$$

From the flow rules, and effective strain function, $d\bar{\epsilon} = -d\epsilon_3$ and $\bar{\sigma} = \sigma_1$, so

$$\frac{d\bar{\sigma}}{\bar{\sigma}} = \frac{3}{2} d\bar{\epsilon} \quad \text{or} \quad \frac{d\bar{\sigma}}{d\bar{\epsilon}} = \frac{3}{2} \bar{\sigma} \tag{4-17}$$

For power-law hardening, instability occurs when

$$\bar{\epsilon} = \frac{2}{3} n \tag{4-18}$$

Regarding pressure, this can be expressed in terms of $\bar{\epsilon}$ by noting that

$$t = t_o e^{\epsilon_3} = t_o e^{-\bar{\epsilon}} \qquad (4\text{-}19)$$

and $\qquad\qquad\qquad r = r_o e^{\epsilon_1} = r_o e^{\bar{\epsilon}/2} \qquad (4\text{-}20)$

Using Eqs. (4-19) and (4-20) and $\bar{\sigma} = K\bar{\epsilon}^n$ in Eq. (4-13),

$$P = 2K\bar{\epsilon}^n \frac{t_o}{r_o} e^{-3/2\bar{\epsilon}} \qquad (4\text{-}21)$$

Figure 4-4 illustrates the variation in pressure with $\bar{\epsilon}$ for a metal where $n = 0.25$. Since the maximum pressure occurs when $\bar{\epsilon} = \frac{2}{3}n$,

$$P_{\max} = 2K\frac{t_o}{r_o}\left(\frac{2}{3}n\right)^n e^{-n} \qquad (4\text{-}22)$$

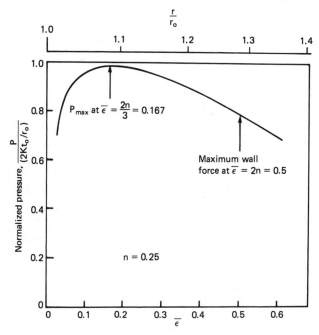

Figure 4-4 Variation of internal pressure with effective strain for a thin-walled sphere. Note that the maximum wall force does not occur at maximum pressure.

4-5 SIGNIFICANCE OF INSTABILITY

Although the condition of instability generally relates to the attainment of a force, pressure, etc., at which the material would continue to deform in a relatively uncontrolled manner, this does not imply that additional and useful uniform deformation always ceases at instability, as is true in a uniaxial tensile test. Reconsider the case of the thin-walled sphere under internal pressure. Although the maximum pressure, hence instability, occurs when $\bar{\epsilon} = \frac{2}{3}n$ as shown in Eq. (4-18), no strain localization or catastrophic failure results;[†] instead the sphere would continue to expand uniformly under a decreasing pressure. Since the walls are subjected to balanced biaxial tension, Eq. (4-12) shows that the *force* in the walls increases until $\bar{\epsilon} = 2n$. Even at this instant, localized deformation still cannot result and cause a sharp neck, as in uniaxial tension, since this would lead to an increase in the local area of the wall. If that were to occur, it would have to be accompanied by a decrease in the local radius of curvature, which, via Eq. (4-13) would lower the wall stress in that region as compared with the remainder of the sphere. These various points are described in Fig. 4-4.

If localized necking does occur in such a process, it must be due to inhomogeneities in the material and/or wall thickness, since these can cause the strain to change from a condition of biaxial tension to plane strain.

4-6 EFFECT OF INHOMOGENEITY ON UNIFORM STRAIN[‡]

Both material properties and dimensions are usually assumed to be homogeneous in the types of analyses just discussed, where under uniaxial tension the maximum uniform true strain was equal to n. Real materials, however, are inhomogeneous, since the cross-sectional diameter will display small variations along the gage section of a tensile test specimen; there may also be variations in grain size, texture, or solute concentration, all of which are inhomogeneities and lead to strength variations along the bar. The effect of these latter type inhomogeneities will be assumed to influence only K in the expression $\bar{\sigma} = K\bar{\epsilon}^n$, so they will act in a manner similar to cross-sectional variations in a bar of homogeneous properties (e.g., grain size, etc.).

The effect of inhomogeneity can be illustrated by considering a tensile test specimen having homogeneous material properties but differing dimensions in two regions, a and b, shown in Fig. 4-5. An inhomogeneity factor, f, is defined as

$$f = \frac{A_{ao}}{A_{bo}}, \quad A_{ao} < A_{bo} \tag{4-23}$$

The regions are assumed to be coupled in series so they must support the same tensile force, i.e. $F_a = F_b$, but may differ in the strain induced. Due to force

[†]This is true even when the pressurizing medium is an ideal gas rather than a fluid. Problem 4-8 addresses this point.

[‡]The basis of this approach was first suggested by Z. Marciniak and K. Kuczynski, *Int. J. Mech. Sci.*, 9, p. 609 (1967).

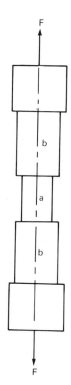

Figure 4-5 Tensile specimen with a dimensional inhomogeneity.

equivalence,

$$A_a \sigma_a = A_b \sigma_b \qquad (4\text{-}24)$$

With power-law hardening, and letting $A_i = A_{io} e^{-\epsilon_i}$, where A_{io} is the initial area at section i, Eq. (4-24) may be altered to

$$A_{ao} e^{-\epsilon_a} K \epsilon_a^n = A_{bo} e^{-\epsilon_b} K \epsilon_b^n \qquad (4\text{-}25)$$

and, using Eq. (4-23), this becomes

$$f \epsilon_a^n e^{-\epsilon_a} = \epsilon_b^n e^{-\epsilon_b} \qquad (4\text{-}26)$$

For given values of f and n, Eq. (4-26) can be solved numerically to give ϵ_b as a function of ϵ_a up to a value of $\epsilon_a = n$ where necking would occur. Figure 4-6 shows such variations for several values of f and a value of n taken as 0.25. Note that as $f \rightarrow 1$ (complete homogeneity), $\epsilon_b \rightarrow \epsilon_a$ up to a limiting strain of 0.25 when necking would occur. This implies a maximum uniform strain outside the neck equal to n as shown by Eq. (4-6). However, as f falls below unity, as expected in actual situations, the strain in region b lags behind that in a and saturates at a level ϵ_b^* which may be considerably less than n. Figure 4-7, again using $n = 0.25$, shows such results, and it is apparent that even

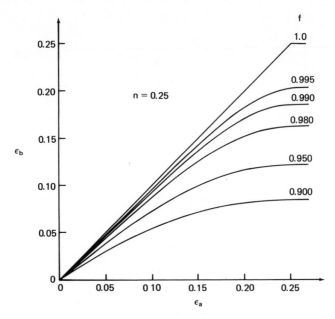

Figure 4-6 Comparison of strains induced in tensile specimens for different inhomogeneity factors, f, assuming $n = 0.25$. Figure 4-5 indicates regions a and b.

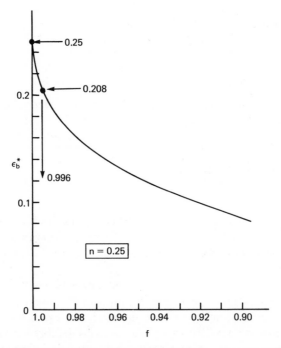

Figure 4-7 Effect of dimensional inhomogeneity factor, f, on the limiting strain outside the neck, ϵ_b^*, during uniaxial tension assuming $n = 0.25$.

small levels of inhomogeneity have appreciable effects. Note that a typical tensile test specimen of 0.505-in. nominal diameter having a maximum diametral variation of 0.001 in. means that $f = 0.996$ and ϵ_b^* is 0.208 instead of 0.250 when f is unity. Figure 4-8 shows the combined influence of n and f on ϵ_b^* and indicates that uniform elongation is strongly dependent upon n as well as f.

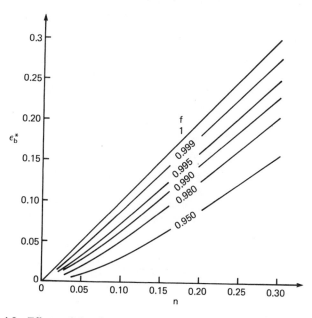

Figure 4-8   Effects of f and n on the limiting strain outside the neck during uniaxial tension. Note that when $f = 1.0$ (no inhomogeneity), $\epsilon_b^* = n$.

PROBLEMS

4-1. If $\bar{\sigma} = K\bar{\epsilon}^n$, the onset of tensile instability occurs when $n = \epsilon_u$. Determine the instability strain as a function of n if
   (a) $\bar{\sigma} = A(B + \bar{\epsilon})^n$ where A and B are constants.
   (b) $\bar{\sigma} = Ae^n$ where e is the nominal strain.

4-2. A rubber balloon displays linear σ-ϵ behavior to fracture and Poisson's ratio is $\frac{1}{2}$. If a balloon of initial diameter d_o is then inflated, what is its diameter, d, when the highest internal pressure is reached?

4-3. Determine the instability strain in terms of n, where $\bar{\sigma} = K\bar{\epsilon}^n$, of a metal that is first subjected to a hydrostatic pressure P, then loaded in uniaxial tension. Note that P is an all-around pressure and remains constant as the tensile load is increased.

4-4. A thin-walled tube with closed ends is pressurized internally and $\bar{\sigma} = 22,000\bar{\epsilon}^{0.25}$ pertains.

(a) At what value of $\bar{\epsilon}$ will instability occur with respect to pressure?

(b) If the tube had an initial diameter of 4 in. and a wall thickness of $\frac{1}{16}$ in., what is the pressure at instability?

4-5. The sketch shows an aluminum tube, whose d_o is 10 in. and t_o is 0.200 in., just fitted over a solid steel rod of 10-in. diameter; the steel may be considered rigid for this situation and friction at the interface is negligible. If $\bar{\sigma} = 25{,}000\bar{\epsilon}^{0.25}$ for the tube and it is loaded as indicated, find $\bar{\epsilon}$ and the force F at instability.

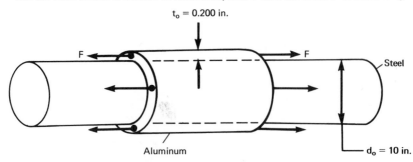

$t_o = 0.200$ in.

Steel

Aluminum

$d_o = 10$ in.

4-6. A metal whose $\bar{\sigma} = 50{,}000\bar{\epsilon}^{0.4}$ undergoes plane-strain tension as shown in the sketch, i.e., $\epsilon_y = 0$.

(a) At the onset of instability determine ϵ_z.

(b) What is the magnitude of $\bar{\sigma}$ at that instant?

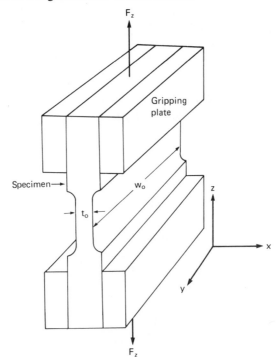

F_z

Gripping plate

Specimen

w_o

z

t_o

x

y

F_z

4-7. A thin-walled tube with closed ends is subjected to an ever-increasing internal pressure. Its behavior is described by $\bar{\sigma} = 100,000\bar{\epsilon}^{0.2}$. At the onset of instability determine the dimensions of the radius, r, and wall thickness, t, in terms of their initial values of r_o and t_o.

4-8. Consider the internal pressurization of a thin-walled sphere by an ideal gas where $PV =$ constant is to be used. One may envision an instability condition when the rate of decrease in pressure with volume, $(-dP/dV)_g$, due to gas expansion is less than the rate of decrease of pressure with volume that the sphere can contain, $(-dP/dV)_{sph}$. When such a condition occurs, a catastrophic expansion would result. If $\bar{\sigma} = K\bar{\epsilon}^n$, determine whether there is such an instability condition and if so, find the necessary value of $\bar{\epsilon}$ as a function of n.

4-9. A tensile bar was machined with a stepped gage section consisting of two regions of different diameters. The initial diameters of the two regions were 2.000 cm and 1.900 cm. After a certain amount of stretching in tension, the diameters of the two regions were measured as 1.893 cm and 1.698 cm respectively. Assuming the tensile strain hardening is described by $\bar{\sigma} = K\epsilon^n$, find n for the material.

4-10. In a rolled sheet, it is not uncommon to find a thickness variation of $\pm 1\%$ from place to place. Consider a sheet 0.030-in. thick (nominally) with $\pm 1\%$ variation of thickness (i.e., some places are $(0.030 - 0.0003)$ in. thick and some places are $(0.030 + 0.0003)$ in. thick). If a tensile specimen cut from the sheet is stretched in tension, how high would n have to be to assure that every part of the sheet is stretched to a strain of at least $\epsilon = 0.20$ before the thinner sections neck?

4-11. Some experts believe that $\bar{\sigma} = A - Be^{-C\epsilon}$ (where A, B, and C are constants and e is the base of natural logarithms) is a better approximation to the true stress-strain curve of metals than a power law.

(a) Using this equation, find the strain at the onset of necking in terms of the constants.

(b) Express the tensile strength, S_u, in terms of these constants.

5

Strain rate
and temperature

5-1 INTRODUCTION

To this point, only the effects of strain on flow stress have been considered; however, flow stress is also dependent upon strain rate and temperature. First, consider the effect of strain rate at constant temperature.

5-2 STRAIN RATE

Usually, flow stress increases with strain rate and the effect at constant strain can be approximated by

$$\sigma = C\dot{\epsilon}^m \tag{5-1}$$

where C is a strength constant that depends upon strain, temperature, and material, and m is the strain-rate sensitivity of the flow stress. For most metals at room temperature, the magnitude of m is quite low (between 0 and 0.03). If the flow stresses, σ_2 and σ_1, at two strain rates, $\dot{\epsilon}_2$ and $\dot{\epsilon}_1$, are compared at the same strain,

$$\frac{\sigma_2}{\sigma_1} = \left(\frac{\dot{\epsilon}_2}{\dot{\epsilon}_1}\right)^m \tag{5-2}$$

or $\ln(\sigma_2/\sigma_1) = m \ln(\dot{\epsilon}_2/\dot{\epsilon}_1)$. If, as is likely at low temperatures, σ_2 is not much greater than σ_1, Eq. (5-2) can be simplified to

$$\frac{\Delta\sigma}{\sigma} \simeq m \ln\frac{\dot{\epsilon}_2}{\dot{\epsilon}_1} = 2.3\,m \log\frac{\dot{\epsilon}_2}{\dot{\epsilon}_1} \tag{5-3}$$

For example, if $m = 0.01$, increasing the strain rate by a factor of 10 would raise the flow stress by only $0.01 \times 2.3 \simeq 2\%$, which illustrates why rate effects are often ignored.

However, rate effects can be important in certain cases. If, for example, one wishes to predict forming loads in wire drawing or sheet rolling (where the strain rates may be as high as 10^4/sec) from data obtained in a laboratory tension test, in which the strain rates may be as low as 10^{-4}/sec, the flow stress should be corrected unless m is very small.

At hot-working temperatures, m typically rises to 0.10 or 0.20, so rate effects are much larger than at room temperature. Under certain circumstances, m-values of 0.5 or higher have been observed in various metals. Ratios of (σ_2/σ_1) calculated from Eq. (5-2) for various levels of $(\dot{\epsilon}_2/\dot{\epsilon}_1)$ and m are shown in Fig. 5-1.

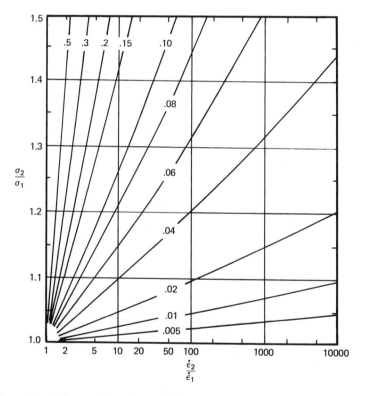

Figure 5-1 Influence of strain rate on flow stress for various levels of strain-rate sensitivity, m, indicated on the curves.

There are two commonly used methods of measuring m. One is to obtain continuous stress-strain curves at several different strain rates and compare the levels of stress at a fixed strain using Eq. (5-2). The other is to make abrupt changes of strain rate during a tension test and use the corresponding level of

$\Delta\sigma$ in Eq. (5-3). These are illustrated in Fig. 5-2. Increased strain rates cause somewhat greater strain hardening, so the use of continuous stress-strain curves yields larger values of m than the second method, which compares the flow stresses for the same structure. The second method has the advantage that several strain-rate changes can be made on one specimen, whereas continuous stress-strain curves require a specimen for each strain rate.

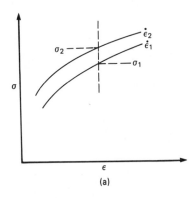

(a)

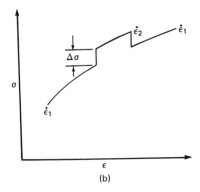

(b)

Figure 5-2 Two methods of determining m. (a) Two continuous stress-strain curves at different strain rates are compared at the same strain and $m = \ln(\sigma_2/\sigma_1)/\ln(\dot{\epsilon}_2/\dot{\epsilon}_1)$. (b) Abrupt strain-rate changes are made during a tension test and $m = (\Delta\sigma/\sigma)/\ln(\dot{\epsilon}_2/\dot{\epsilon}_1)$.

Strain-rate sensitivity is also temperature dependent; Fig. 5-3 shows data for a number of metals obtained from continuous constant strain-rate tests. Below $T/T_m = \frac{1}{2}$, (T/T_m is the ratio of testing temperature to melting point on an absolute scale), the rate sensitivity is low but it climbs rapidly for $T > T_m/2$.

More detailed data for aluminum alloys are given in Fig. 5-4. Although the definition of m in this figure is based upon shear stress and strain rate, it is equivalent to the definition derived from Eq. (5-1).

For these and many other alloys there is a minimum in m near room temperature and, as indicated, negative m-values are sometimes found. At low strain rates, solutes segregate to dislocations; this lowers their energy so that the forces required to move the dislocations are higher than those required for

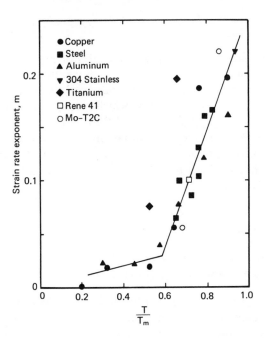

Figure 5-3 Variation of the strain-rate sensitivity of different materials with homologous temperature, T/T_m. Adapted from F. W. Boulger, *DMIC Report 226*, Battelle Mem. Inst. (1966), pp. 13–37.

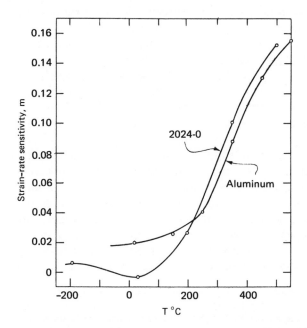

Figure 5-4 Temperature dependence of the strain-rate sensitivities of 2024 and pure aluminum. From D. S. Fields and W. A. Backofen, *Trans. ASM*, 51 (1959), pp. 946–60.

solute-free dislocations. At increased strain rates or lower temperatures, however, dislocations move faster than solute atoms can diffuse, so the dislocations are relatively solute-free and the drag is minimized. A negative rate sensitivity tends to localize flow in a narrow region which propagates along a tensile specimen as a Lüders band. By localization of flow in a narrow band, the deforming material experiences a higher strain rate and therefore a lower flow stress.

The higher values of rate sensitivity at elevated temperatures are attributed to the increased rate of thermally activated processes such as dislocation climb and grain-boundary sliding.

5-3 SUPERPLASTICITY

As mentioned earlier, rate sensitivities of 0.5 or greater can sometimes be encountered. The necessary conditions are:

1. An extremely fine grain size (a few microns or less), with generally uniform and equiaxed grain structure,
2. High temperatures ($T > 0.4T_m$), and
3. Low strain rates† (10^{-2}/sec or lower).

In addition to the above, the stability of microstructure during deformation is desired. Under these conditions, extremely high elongations and very low flow stresses are observed in tension tests, so the term *superplasticity* has been used to describe these effects. An excellent review of superplasticity and the mechanisms and characteristics are given by Edington et al.‡

There are two useful aspects of superplasticity. One is the extremely low flow stresses at useful working temperatures which permit creep forging of intricately shaped parts and reproduction of fine detail. The other is the extreme tensile elongations which permit sheet parts of great depth to be formed with simple tooling.

One application of superplasticity has been in the production of titanium alloy aircraft panels. An example of such a panel in cross section is shown in Fig. 5-5. Using superplastic forming with concurrent diffusion bonding (SPF/DB), three sheets are first diffusion bonded at specific locations, bonding elsewhere being prevented by painting the sheets with an inert ceramic. Then the unbonded channels are pressurized internally with argon until the skin pushes against tools which control the outer shape. Superplastic behavior is required to obtain the necessary elongation of the interior core sheet ligaments, which

---

†The fact that m itself is rate dependent indicates that Eq. (5-1) is not a true description of the rate effect. Nevertheless, expressed as $m = (\partial \ln \sigma / \partial \ln \dot{\epsilon})_{\epsilon,T}$ it is always a useful index of the strain-rate sensitivity of the flow stress and is a convenient basis for describing and analyzing strain-rate effects.

‡J. W. Edington, K. N. Melton, and C. P. Cutler, *Progress in Materials Sci.*, 21 (1976), pp. 63–170.

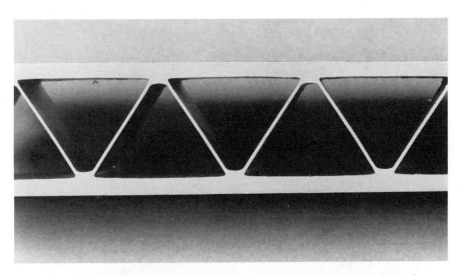

Figure 5-5 Aircraft panel made from a titanium alloy by diffusion bonding and superplastic expansion by internal pressure. Courtesy of Rockwell International Corp.

Figure 5-6 Complex sheet-metal part of Zn-22%Al made by superplastic forming. Courtesy of D. S. Fields, IBM Corp.

is over 100% in the example shown. Figure 5-6 shows an example of a deep part made by superplastic forming of Zn-22%Al sheet. A tensile bar elongated 1950% (a 19-fold increase in length!) is illustrated in Fig. 5-7. To appreciate this, one must realize that a tensile elongation over 50% is usually considered large. The tensile elongations for a number of materials are plotted in Fig. 5-8 as a function of m. High m-values promote large elongations by preventing localization of the deformation as a sharp neck. This can be seen clearly in the following example.

Consider a bar that starts to neck. If the cross-sectional area in the neck

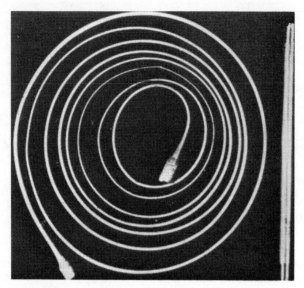

Figure 5-7 Superplastically extended tensile bar of Bi-Sn eutectic. From C. E. Pearson, *J. Inst. Metals*, 54 (1934), p. 111.

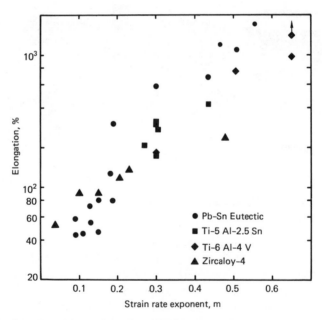

Figure 5-8 Correlation of tensile elongation with strain-rate sensitivity. Data from D. Lee and W. A. Backofen, *TMS-AIME*, 239 (1967), p. 1034; and D. H. Avery and W. A. Backofen, *Trans. Q. ASM*, 58 (1965), pp. 551–62.

is A_n and outside the neck is A_u, $F = \sigma_u A_u = \sigma_n A_n$ or $\sigma_u/\sigma_n = A_n/A_u$. From Eq. (5-2),

$$\frac{\dot{\epsilon}_u}{\dot{\epsilon}_n} = \left(\frac{\sigma_u}{\sigma_n}\right)^{1/m} = \left(\frac{A_n}{A_u}\right)^{1/m} \tag{5-4}$$

Since $A_n < A_u$, if m is low, the strain rate outside the neck will become negligibly low. For example, let the neck region have a cross-sectional area of 90% of that outside the neck. If $m = 0.02$, $\dot{\epsilon}_u/\dot{\epsilon}_n = (0.9)^{50} = 5 \times 10^{-3}$. If, however, $m = 0.5$, then $\dot{\epsilon}_u/\dot{\epsilon}_n = (0.9)^2 = 0.81$, so that although the unnecked region deforms slower than the neck, its rate of straining is still rather large.

For a more complete treatment, we will reconsider the tension test in Chap. 4 of a bar of inhomogeneous cross section, but now we will neglect strain hardening and assume that $\sigma = C\dot{\epsilon}^m$. Again, the bar is divided into two regions, one with an initial cross section of A_{bo} and the other $A_{ao} = fA_{bo}$. As before, the forces must balance, so $\sigma_b A_b = \sigma_a A_a$. Substituting $A_i = A_{io}e^{-\epsilon_i}$ and $\sigma_i = C\dot{\epsilon}_i^m$,

$$A_{bo}e^{-\epsilon_b}\dot{\epsilon}_b^m = A_{ao}e^{-\epsilon_a}\dot{\epsilon}_a^m \tag{5-5}$$

(where ϵ_a and ϵ_b are respectively the strains in the reduced and unreduced sections), or

$$e^{-\epsilon_b}\left(\frac{d\epsilon_b}{dt}\right)^m = fe^{-\epsilon_a}\left(\frac{d\epsilon_a}{dt}\right)^m \tag{5-6}$$

Raising both sides to the $(1/m)$ power and cancelling dt

$$\int_o^{\epsilon_b} e^{-\epsilon_b/m}\,d\epsilon_b = \int_o^{\epsilon_a} f^{1/m}e^{-\epsilon_a/m}\,d\epsilon_a \tag{5-7}$$

Integrating gives

$$e^{-\epsilon_b/m} - 1 = f^{1/m}(e^{-\epsilon_a/m} - 1) \tag{5-8}$$

Numerical solutions of ϵ_b as a function of ϵ_a for $f = 0.98$ and several levels of m are shown in Fig. 5-9. At low levels of m (or low values of f), ϵ_b tends to saturate early and approaches a limiting strain ϵ_b^* at moderate levels of ϵ_a, but with higher m-values, saturation of ϵ_b is much delayed, i.e., localization of strain in the reduced section (or the onset of a sharp neck) is postponed. Thus, the conditions that promote high m-values also promote high failure strains in the necked region. Letting $\epsilon_a \rightarrow \infty$ in Eq. (5-8) will not cause great error and will provide limiting values for ϵ_b^*. With this condition,

$$\epsilon_b^* = -m\ln(1 - f^{1/m}). \tag{5-9}$$

In Fig. 5-10, values of ϵ_b^* calculated from Eq. (5-9) are plotted against m for various levels of f. The values of tensile elongation corresponding to ϵ_b^* are indicated on the right margin. It is now clear why large elongations are observed

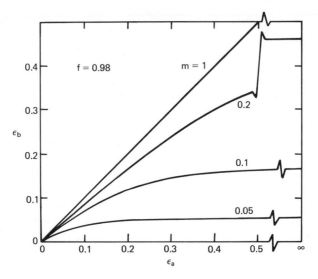

Figure 5-9 Relative strains in unreduced, ϵ_b, and reduced, ϵ_a, sections of a stepped tensile specimen for various levels of m, assuming no strain hardening.

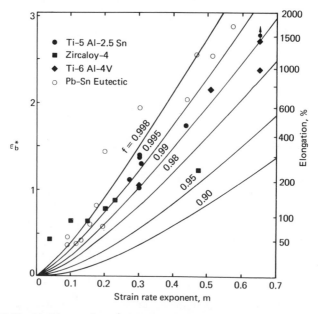

Figure 5-10 Limiting strains, ϵ_b^*, in unreduced sections of stepped tensile specimens as a function of m and f. Values of percent elongation corresponding to ϵ_b^* are indicated on the right together with data from Fig. 5-8.

under superplastic conditions. The data in Fig. 5-8, replotted here, suggest an inhomogeneity factor of about 0.99 to 0.998 (for a round bar 0.250-in. diameter, a diameter variation of 0.0005 in. corresponds to $f = 0.996$). The general agreement is perhaps fortuitous considering the assumptions and simplifications. The values of ϵ_b^* are the strains away from the neck, so the total elongation would be even higher than indicated here. On the other hand, strains in the neck, ϵ_a, are not infinite, so the ϵ_b^* values corresponding to realistic maximum values of ϵ_a will be lower than those used in Fig. 5-10. Also, the experimental values are affected by difficulties in maintaining constant temperature over the length of the bar as well as a constant strain rate in the deforming section. Nevertheless, the agreement between theory and experiments is striking.

Figure 5-11 shows the dependence of flow stress of the Al-Cu eutectic alloy at 520°C upon strain rate. The effect of strain is not important here because

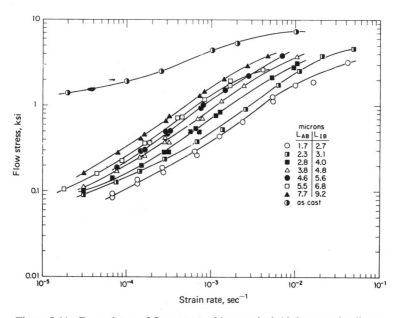

Figure 5-11   Dependence of flow stress of hot-worked Al-Cu eutectic alloy on strain rate at 520°C. The curves are for different grain sizes. L_{AB} is mean free path within the grains of κ and $CuAl_2$ phases and L_{IB} the mean free path between interphase boundaries. From D. A. Holt and W. A. Backofen, *Trans. Q. ASM*, 59 (1966), p. 755.

strain hardening is negligible at this high temperature. Figure 5-12 shows the corresponding value of m as a function of strain rate. At the higher strain rates, m is typical of thermally activated slip. At lower strain rates, deformation mechanisms other than slip prevail. Here there are two schools of thought.

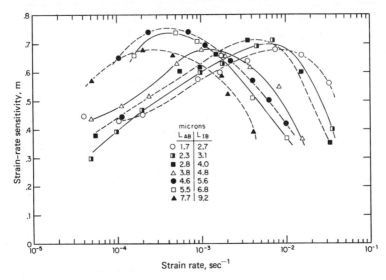

Figure 5-12 Variation of strain-rate sensitivity, $m = (\partial\ln\sigma/\partial\ln\dot\epsilon)\epsilon$, T for the Al-Cu eutectic at 520°C. Compare with Fig. 5-11. Note that the strain rate for peak m increases with decreasing grain size. From D. A. Holt and W. A. Backofen, *ibid*.

One[†][‡] maintains that deformation occurs primarily by diffusional creep with *vacancies* migrating from grain boundaries normal to the tensile axis to those parallel to it (i.e., diffusion of *atoms* from boundaries parallel to the tensile axis to boundaries perpendicular to it). This diffusion causes the grains to elongate in the tensile direction and contract laterally. Whether diffusion is through the lattice or along grain-boundary paths, the strain rate should be proportional to the applied stress and inversely related to the grain size. If diffusion were the only mechanism, m would equal one (Newtonian viscosity) but it is lowered because of the slip contribution to the overall strain. The other school[§] attributes the high rate sensitivity to the role of grain-boundary sliding (shearing on grain boundaries). Although grain-boundary sliding alone would be viscous ($m = 1$), it must be accompanied by another mechanism to accommodate compatability at triple points where the grains meet. Either slip or diffusion could serve as the accommodating mechanism. Both models explain the need for a very fine grain size, high temperature, and low strain rate, but

[†]W. A. Backofen, *Deformation Processing* (Reading, Mass.: Addison-Wesley, 1972), pp. 217–20.

[‡]A. H. Cottrell, *The Mechanical Properties of Matter* (New York: John Wiley, 1964), p. 202.

[§]M. F. Ashby and R. A. Verrall, *Acta Met.*, 21 (1973), p. 149.

the diffusional-creep model does not explain why the grains remain equiaxed after large deformations.

One practical problem is the tendency for grain growth during super-plastic deformation because of the high temperatures and long times dictated by the low strain rates, and because of the superplastic deformation itself.† Such grain growth would lower the m-value,‡ shift peak m-values to still lower strain rates, and increase the flow-stress values, and thus deteriorate the overall superplastic formability. Because a large amount of second phase markedly retards grain growth, most superplastic alloys have two-phase microstructures. Among the alloys which exhibit superplasticity are Zn-22%Al (eutectoid), Sn-40% Pb, Sn-5%Bi, Al-33%Cu (eutectics), Cu-10%Al, Zircaloy and several titanium alloys (two-phase α-β), a range of steels (for example, AISI 1340 and 0.2 to 1.0% C-steels) and some superalloys (γ-γ' with carbides).

5-4 COMBINED STRAIN AND STRAIN-RATE EFFECTS

Even the low values of m at room temperature can be of importance in deter-mining uniform elongation. In Chap. 4 it was shown that the uniform elongation in a tension test was controlled by the strain-hardening exponent and the inhomogeneity factor f, but strain-rate effects were neglected. Now, reconsider the tension test on an inhomogeneous specimen with two regions of initial cross-sectional areas A_{bo} and A_{ao} ($= f A_{bo}$). Assume now that the material work hardens and is also rate sensitive, so that the flow stress is given by

$$\sigma = c' \epsilon^n \dot{\epsilon}^m \qquad (5\text{-}10)$$

Substituting $A_i = A_{io} e^{-\epsilon_i}$ and $\sigma = c' \epsilon_i^n \dot{\epsilon}_i^m$ into a force balance, $A_b \sigma_b = A_a \sigma_a$, results in

$$A_{bo} e^{-\epsilon_b} \epsilon_b^n \dot{\epsilon}_b^m = A_{ao} e^{-\epsilon_a} \epsilon_a^n \dot{\epsilon}_a^m$$

Following the procedure that produced Eq. (5-7),

$$\int_0^{\epsilon_b} e^{-\epsilon_b/m} \epsilon_b^{n/m} d\epsilon_b = f^{1/m} \int_0^{\epsilon_a} e^{-\epsilon_a/m} \epsilon_a^{n/m} d\epsilon_a \qquad (5\text{-}11)$$

The results of integration and numerical evaluation are shown in Fig. 5-13, where ϵ_b is plotted as a function of ϵ_a for $n = 0.2$, $f = 0.98$, and several levels of m. It is apparent that even quite low levels of m play a significant role in controlling the strains reached in the unnecked region of the bar.

---

†S. P. Agrawal and E. D. Weisert, in *7th North American Metalworking Research Conference*, Soc. Man. Engrs., Dearborn (1979), pp. 197–204.

‡The low m-values often observed at very low strain rates have been attributed in part to grain coarsening during these experiments.

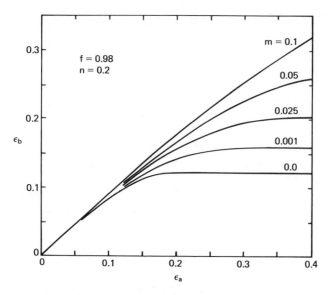

Figure 5-13 Comparison of the strains in reduced and unreduced sections of a tensile specimen with an inhomogeneity factor of $f = 0.98$ calculated from Eq. (5-11) with $n = 0.20$. Even at relatively low levels, m significantly influences ϵ_b.

5-5 ALTERNATIVE DESCRIPTION OF STRAIN-RATE DEPENDENCE

For steels, Eq. (5-1) may not be the best description of the strain-rate dependence of flow stress. There is considerable evidence that the strain-rate exponent, m, decreases as the steel is strain hardened and that it is lower for high strength steels than for weaker steels. The data of Fig. 5-14 show that m is inversely proportional to the flow stress, σ. This suggests that a better description is

$$\frac{d\sigma}{d(\ln \dot{\epsilon})} = m' \tag{5-12}$$

or

$$\sigma = m' \ln \frac{\dot{\epsilon}}{\dot{\epsilon}_o} + C \tag{5-13}$$

where m' is the new rate-sensitivity constant and C is the value of σ at $\dot{\epsilon} = \dot{\epsilon}_o$ (C and $\dot{\epsilon}_o$ are not independent constants; if $\dot{\epsilon}_o$ is taken as unity, $C = \sigma$ for $\dot{\epsilon} = 1$).

Since C must depend upon strain, Eq. (5-13) indicates that the contributions of strain and strain rate to flow stress are *additive* rather than multiplicative as in Eq. (5-10). Another indication supporting this postulate is the observation that the strain-hardening exponent, n, for steels decreases at high strain rates; see Fig. 5-15.

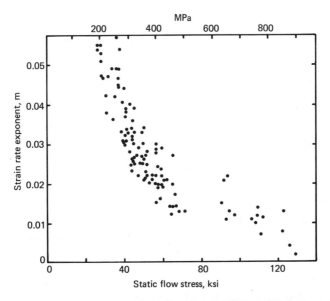

Figure 5-14 Effect of stress level on strain-rate sensitivity of steels. Adapted from A. Saxena and D. A. Chatfield, *SAE Paper 760209* (1976).

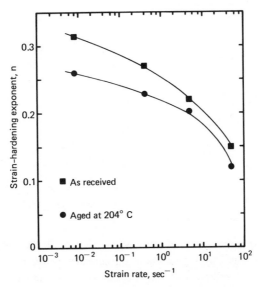

Figure 5-15 Dependence of the strain-hardening exponent, n, upon strain rate for steels. Adapted from A. Saxena and D. A. Chatfield, *ibid.*

When C in Eq. (5-13) is $K' \epsilon^{n'}$,

$$\sigma = m' \ln \frac{\dot{\epsilon}}{\dot{\epsilon}_o} + K' \epsilon^{n'} \qquad (5\text{-}14)$$

The usual exponent n in $\sigma = K\epsilon^n$ can be expressed as

$$n = \frac{\epsilon \, d\sigma}{\sigma \, d\epsilon} \qquad (5\text{-}15)$$

Then, evaluating n in terms of Eq. 5-14,

$$n = n' \left/ \left[\frac{m' \ln (\dot{\epsilon}/\dot{\epsilon}_o)}{K' \epsilon^{n'}} + 1 \right] \right. \qquad (5\text{-}16)$$

so if m' and n' are truly constant, n would appear to decrease with strain rate.

Data for fcc metals, however, indicate that Eq. (5-10) is better than Eq. (5-14). Even with steels, the simpler mathematics of Eq. (5-10) justify its use as long as m and n are chosen at representative stress levels.

5-6 TEMPERATURE DEPENDENCE OF FLOW STRESS

At elevated temperatures the rate of strain hardening falls rapidly in most metals with an increase in temperature, as shown in Fig. 5-16. The flow stress and tensile strength, measured at constant strain and strain rate, also drop

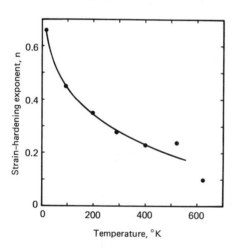

Figure 5-16 Decrease of the strain-hardening exponent, n, of pure aluminum with temperature. Adapted from R. P. Carreker and W. R. Hibbard, Jr., *Trans. TMS-AIME*, 209 (1957), pp. 1157–63.

with increasing temperature as illustrated in Fig. 5-17. However, the drop is not always continuous; often there is a temperature range over which the flow stress is only slightly temperature dependent or in some cases even increases slightly with temperature. The temperature dependence of flow stress is closely related to its strain-rate dependence. Decreasing the strain rate has the same effect on flow stress as raising the temperature, as indicated schematically in

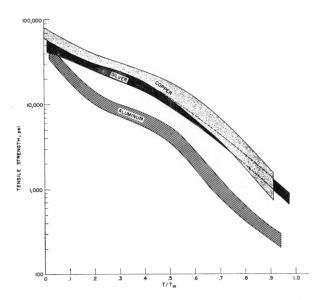

Figure 5-17 Decrease of tensile strength of pure copper, silver, and aluminum with homologous temperature. From R. P. Carrecker and W. R. Hibbard, Jr., *ibid*.

Fig. 5-18. Here it is clear that at a given temperature the strain-rate dependence is related to the slope of the σ versus T curve; where σ increases with T, m must be negative.

The simplest quantitative treatment of temperature dependence is that of Zener and Hollomon[†] who argued that plastic straining could be treated as a rate process using the Arrhenius rate law,[‡] rate $\propto \exp(-Q/RT)$, which has been successfully applied to many rate processes. They proposed that

$$\dot{\epsilon} = A e^{-Q/RT} \tag{5-17}$$

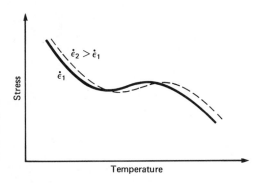

Figure 5-18 Schematic plot showing the temperature dependence of flow stress for some alloys. In the temperature region where flow stress increases with temperature, the strain-rate sensitivity is negative.

[†]C. Zener and H. H. Hollomon, *J. Appl. Phys.*, 15 (1944), pp. 22–32.

[‡]S. Arrhenius, *Z. Phys. Chem.*, 4 (1889), p. 226.

where Q is an activation energy, T the absolute temperature, and R the gas constant. Here the constant of proportionality, A, is both stress and strain dependent. At constant strain, A is a function of stress alone $[A = A(\sigma)]$, so Eq. (5-17) can be written as

$$A(\sigma) = \dot{\epsilon}e^{Q/RT} \tag{5-18}$$

or more simply as

$$\sigma = f(z) \tag{5-19}$$

where the Zener-Hollomon parameter $Z = \dot{\epsilon}e^{Q/RT}$. This development predicts that if the strain rate to produce a given stress at a given temperature is plotted on a logarithmic scale against $1/T$, a straight line should result with a slope of $-Q/R$. Figure 5-19 shows such a plot for 2024-O aluminum.

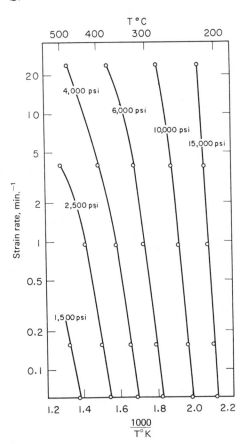

Figure 5-19 Strain rate and temperature combination for various levels of flow stress. From D. S. Fields and W. A. Backofen, *ibid.*

Correlations of this type are very useful in relating temperature and strain-rate effects, particularly in the high-temperature range. However, such correlations may break down if applied over too large a range of temperatures, strains, or strain rates. One reason is that the rate-controlling process, and hence Q, may change with temperature or strain. Another is connected with the original formulation of the Arrhenius rate law in which it was supposed that thermal fluctuations alone overcome an activation barrier, whereas in plastic deformation, the applied stress acts together with thermal fluctuations in overcoming the barriers as indicated in the following development.

Consider an activation barrier for the rate-controlling process, as in Fig. 5-20. The process may be cross slip, dislocation climb, etc. Ignoring the details,

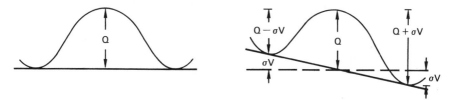

Figure 5-20 Schematic illustration of an activation barrier for slip and the effect of applied stress on skewing the barrier.

assume that the dislocation moves from left to right. In the absence of applied stress, the activation barrier has a height Q and the rate of overcoming this barrier would be proportional to $\exp(-Q/RT)$. However, unless the position at the right is more stable, i.e., has a lower energy than the position on the left, the rate of overcoming the barrier from right to left would be exactly equal to that in overcoming it from left to right, so there would be no net dislocation movement. With an applied stress, σ, the energy on the left is raised by σV, where V is a constant with dimensions of volume, and on the right the energy is lowered by σV. Thus the rate from left to right is proportional to $\exp[-(Q - \sigma V)/RT]$ and from right to left the rate is proportional to $\exp[-(Q + \sigma V)/RT]$. The net strain rate then is

$$\dot{\epsilon} = C\{\exp[-(Q - \sigma V)/RT] - \exp[-(Q + \sigma V)/RT]\}$$
$$= C\exp(-Q/RT)[\exp(\sigma V/RT) - \exp(-\sigma V/RT)]$$
$$= 2C\exp(-Q/RT)\sinh(\sigma V/RT) \qquad (5\text{-}20)$$

To accommodate data better, and for some theoretical reasons, a modification of Eq. (5-20) has been suggested.[†‡] It is:

$$\dot{\epsilon} = A[\sinh(\alpha\sigma)]^{1/m}\exp(-Q/RT) \qquad (5\text{-}21)$$

Steady-state creep data over many orders of magnitude of strain rate correlate very well with Eq. (5-21), as shown in Fig. 5-21.

It should be noted that if $\alpha\sigma \ll 1$, $\sinh(\alpha\sigma) \approx \alpha\sigma$ so Eq. (5-21) reduces to

$$\dot{\epsilon} = A\exp(-Q/RT) \cdot (\alpha\sigma)^{1/m}$$

†F. Garofalo, *TMS-AIME*, 227, p. 251 (1963).

‡J. J. Jonas, C. M. Sellers, and W. J. McG Tegart, *Met. Rev.*, 14, p. 1 (1969).

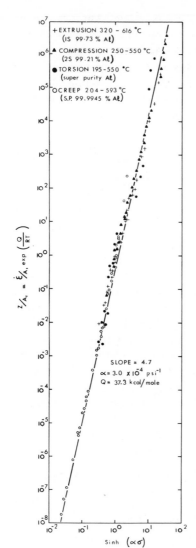

Figure 5-21 Plot of Zener-Hollomon parameter versus flow-stress data for aluminum showing the validity of the hyperbolic sine relationship, Eq. (5-21). From J. J. Jonas, *Trans. Q. ASM*, 62 (1969), pp. 300–3.

or

$$\sigma = A'\dot{\epsilon}^m \exp(mQ/RT) \qquad (5\text{-}22)$$

$$\sigma = A'Z^m$$

which is consistent with both the Zener-Hollomon development, Eq. (5-19) and the power-law expression, Eq. (5-1).

Since $\sinh(x) \to e^x/2$ for $x \gg 1$, at low temperatures and high stresses Eq. (5-21) reduces to

$$\dot{\epsilon} = C \exp(\alpha'\sigma - Q/RT) \qquad (5\text{-}23)$$

but now strain hardening becomes important so C and α' are both strain and temperature dependent. Equation (5-23) reduces to

$$\sigma = C + m' \ln \dot{\epsilon} \qquad (5\text{-}24)$$

which is consistent with Eq. (5-13) and explains the often observed breakdown in the power-law strain-rate dependence at low temperatures and high strain rates.

5-7 HOT WORKING

Hot working takes advantage of the decrease in flow stress at high temperatures to lower tool forces and, consequently, equipment size and power requirements. Hot working is often defined as working above the recrystallization temperature so that the work metal recrystallizes as it deforms. However, this is an over-simplified view. The strain rates of many metal-working processes are so high that there is not time for recrystallization to occur during deformation. Rather, recrystallization may occur in the time period between repeated operations, as in forging and multiple-stand rolling, or after the deformation is complete and the material begins to cool to room temperature. The high temperature does, however, lower the flow stress whether recrystallization occurs during the deformation or not. Furthermore, the resultant product is in an annealed state.

In addition to lowering the flow stress, the elevated temperature during hot working has several undesirable effects. Among them are:

1. Lubrication is more difficult. Although viscous glasses are often used in hot extrusions, much hot working is done without lubrication.
2. The work metal tends to oxidize. Scaling of steel and copper alloys causes loss of metal and roughened surfaces. While processing under inert atmosphere is possible, it is prohibitively expensive and is avoided except in the case of very reactive metals.
3. Tool life is shortened because of heating, the presence of abrasive scales, and the lack of lubrication. Sometimes scale breakers are employed and rolls are cooled by water spray to minimize tool damage.
4. Poor surface finish and loss of precise gauge control result from the lack of adequate lubrication, oxide scales, and roughened tools.
5. The lack of work hardening is undesirable where the strength level of a cold-worked product is needed.

Because of these limitations, it is common to hot roll steel to about 0.10-in. thickness to take advantage of the decreased flow stress at high temperature. The hot-rolled product is then pickled to remove scale, and further rolling is done cold to insure good surface finish and optimum mechanical properties.

5-8 TEMPERATURE RISE DURING DEFORMATION

The temperature of the metal rises during plastic deformation because of the heat generated by mechanical work. The mechanical energy per volume, w, expended in deformation is equal to the area under the stress-strain curve

$$w = \int_{o}^{\bar{\epsilon}} \bar{\sigma} \, d\bar{\epsilon} \tag{5-25}$$

Only a small fraction of this energy is stored (principally as dislocations and vacancies). This fraction drops from about 5% initially to 1 or 2% at high strains. The rest is released as heat. If the deformation is adiabatic, i.e., no heat transfer to the surroundings, the temperature rise is given by

$$\Delta T = \frac{\alpha \int \bar{\sigma} \, d\bar{\epsilon}}{\rho C} = \frac{\alpha \bar{\sigma}_a \bar{\epsilon}}{\rho C} \tag{5-26}$$

where $\bar{\sigma}_a$ is the average value of σ over the strain interval o to ϵ, ρ is the density, C is the mass heat capacity, and α is the fraction of energy stored (~ 0.98).

Example 5-1

Calculate the temperature rise in a high-strength steel that is adiabatically deformed to a strain of 1.0. Pertinent data are: $\rho = 7.87 \times 10^3$ kg/m³, $\sigma_a = 800$ MPa, $C = 0.46 \times 10^3$ J/kg°C.

Solution. Substituting in Eq. (5-26) and taking $\alpha = 1$

$$\Delta T = \frac{800 \times 10^6 \times 1.0}{7.87 \times 10^3 \times 0.46 \times 10^3} = 221°C$$

Although both σ_a and ϵ were high in this example, it illustrates the possibility of large temperature rises during plastic deformation.

Any temperature increase causes the flow stress to drop. One effect is that at low strain rates, where heat can be transferred to surroundings, there is less thermal softening than at high strain rates. This partially compensates for the strain-rate effect on flow stress and can lead to an apparent decrease in the strain-rate sensitivity at high strains, when m is derived by comparing continuous stress-strain curves.

Another effect of the thermal softening is that it can act to localize flow in narrow bands. If one region or band deforms more than another, the greater heating may lower the flow stress in this region, causing even more concentration of flow and more local heating in this region. An example of this is shown in Fig. 5-22, where in the upsetting of a steel bolt head, localized flow along narrow bands raised the temperature sufficiently to cause the bands to transform to austenite. After the deformation, these bands were quenched to martensite by the surrounding material. Similar localized heating, reported in punching holes in armor plate, can lead to sudden drops in the punching force. Such extreme localization is encouraged by conditions of high flow stresses and strains [increasing ΔT in Eq. (5-26)], low rates of work hardening (high work hardening would tend to prevent localization), and workpiece-tool geometry that encourages deformation along certain discrete planes.

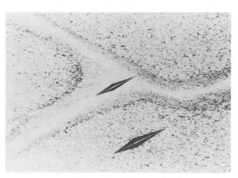

Figure 5-22   Flow pattern and microstructure of a bolt head of quenched and tempered 8640 steel after cold upsetting at a high velocity. The light bands of untempered martensite show a higher hardness, as indicated by the Knoop indentations at the right. From F. W. Boulger, *ibid.*

PROBLEMS

5-1. To achieve weight savings in automobiles, low-carbon steels are being replaced by HSLA steels in some components. The higher yield strengths of the HSLA steels permit the use of thinner gauges. In laboratory tension tests (at a strain rate of about 10^{-3}/sec) one grade of HSLA steel has a yield strength of 60,000 psi with a strain-rate exponent of $m = 0.01$, and for a low-carbon steel $Y = 35,000$ psi and $m = 0.03$.

Calculate the percent weight saving that could be achieved by substitution of the HSLA steel for the low-carbon steel, assuming the component is redesigned to carry the same tensile load without yielding:

(a) at laboratory strain rates;

(b) in an automobile crash in which the strain rate is estimated to be about 10^4/sec.

5-2. The thickness of a rolled sheet is found to vary from 0.0322 to 0.0318 in., depending upon where the measurements are made, so strip tensile specimens cut from the sheet show similar variation in cross section.

(a) For a material with $n = 0.15$, $m = 0$, what will be the strain in the thicker region when the thinner region necks?

(b) Find the strains in the thicker region if $m = 0.05$ and $n = 0$ when the strain in the thinner region reaches 0.5 and ∞.

5-3. (a) What value of tensile elongation would be required in the ligaments of Fig. 5-5 if the ligaments were to make an angle of 75° with the horizontal?

(b) Assuming $f = 0.98$ and $n = 0$, what value of m would be required if the thickness variation along the ligament is to be held to 20%?

5-4. Find the value of m' in Eq. (5-12) which best fits the data in Fig. 5-14 for steels.

5-5. From the data in Fig. 5-19, estimate the value of Q in Eq. (5-18) and m in Eq. (5-1) for aluminum at 400°C.

5-6. Estimate the total elongation of a superplastic tensile bar ($n = 0$) if
   (a) $f = 0.98, m = 0.5$
   (b) $f = 0.75, m = 0.8$

5-7. Estimate the local shear strain in the shear bands of Fig. 5-22 to explain the formation of untempered martensite. Assume that the strength level is 250,000 psi, no work hardening occurs, and adiabatic conditions prevail.

5-8. In superplastic forming it is often necessary to control the strain rate. Consider the forming of a sheet into a hemispherical dome by clamping it over a circular hole and bulging it with gas pressure.
   (a) Describe qualitatively how the gas pressure should be varied during the cycle if a constant strain rate is to be maintained at the dome.
   (b) Compare the levels of gas pressure to form hemispherical domes of 2-in. dia. and 20-in. dia. from sheet of the same thickness at the same strain rate.

5-9. **(a)** During a creep experiment under a constant stress, the temperature is suddenly increased from 290 to 300°C and the rate of creep is found to double. What is the apparent activation energy for creep?
   (b) During a tension test, the rate of straining is suddenly increased by a factor of 8. The level of the stress-strain curve correspondingly rises 1.8%. What is the value of m?

6

Ideal work
or uniform energy

6-1 INTRODUCTION

One concern regarding forming operations is a prediction of the externally
applied loads needed to cause the metal to flow and deform to the shape
desired. Because of uncertainties introduced by frictional effects, non-homogeneous deformation, and the true manner by which strain hardening occurs
during complex deformations, exact values of force requirements are seldom
predictable. As a recourse, several techniques have been developed which,
using pertinent assumptions, enable an approximation of load requirements.
To varying degrees, each of the uncertainties mentioned above is introduced
into these analytical methods, thereby permitting an estimation of the deformation forces, the constraining forces, and the manner in which metal flow
occurs. Unfortunately, these techniques do not, in general, provide a means to
predict the mechanical properties of the deformed material, the maximum possible deformation up to fracture, or any variation in frictional effects during
deformation.

6-2 IDEAL WORK OR UNIFORM ENERGY

The simplest method for force prediction is a work or energy balance. The
external work is equated to the energy consumed in deforming the workpiece.
The process is assumed to be ideal in the sense that external work is completely
utilized to cause deformation only. Effects of friction and non-homogeneous
deformation are ignored.

To employ this technique, it is necessary to envision an ideal process, whether practical or not, that produces the desired shape change by homogeneous deformation. As an example, axisymmetric extrusion or wire (rod) drawing causes a bar of some starting diameter to be reduced to some smaller diameter; the ideal process that matches this geometric change is a tension test. As discussed earlier, to reduce a bar of area A_o to area A under tensile loading requires work per unit volume equal to the area under the true stress-true strain curve within the strain limits of concern. Since this is an *ideal* process, there results

$$w_i = \int_0^{\bar{\epsilon}} \bar{\sigma} d\bar{\epsilon} \tag{6-1}$$

where $\bar{\epsilon} = \ln(A_o/A)$. Assuming power-law hardening,

$$w_i = \frac{K\bar{\epsilon}^{n+1}}{n+1} \tag{6-2}$$

and if other forms of work hardening are more appropriate, they would be used in Eq. (6-1). What is essential is a knowledge of the dependence of $\bar{\sigma}$ on $\bar{\epsilon}$ and not whether the proposed area reduction could be actually accomplished in a tension test without necking.

Example 6-1

Consider a metal whose strain-hardening behavior follows $\bar{\sigma} = 25{,}000\bar{\epsilon}^{0.25}$. If an annealed bar of this metal is pulled in a tension test from a starting diameter of 12.7 mm to a diameter of 11.5 mm, what is the work per unit volume required?

Solution.   This is an ideal process so Eq. (6-2) may be used, however, one must be aware of units. K in the above strain-hardening equation has units of psi (these numbers are typical of an annealed 1100 aluminium).

$$\bar{\epsilon} = 2\ln\frac{12.7}{11.5} = 0.199$$

so $w_i = 25000(.199)^{1.25}/1.25 = 2650$ in. lb/in³.

For a forging operation, a frictionless compression test might be viewed as the ideal process, whereas in flat rolling, a plane-strain tension test could serve as such a model. Regardless of the operation, w_i provides the basis for analysis and the equivalence of operations that produce the same geometric change of shape is the fundamental requirement.

On occasion, a mean flow stress, Y_m, is sometimes used over the range of homogeneous strain involved such that

$$w_i = Y_m\bar{\epsilon} \tag{6-3}$$

and by comparing Eqs. (6-2) and (6-3),

$$Y_m = \frac{K\bar{\epsilon}^n}{n+1} = \frac{\bar{\sigma}}{n+1} \tag{6-4}$$

6-3 EXTRUSION AND ROD DRAWING

Figure 6-1 illustrates a direct or forward extrusion process where a billet of initial diameter D_o is forced through a converging die to exit with diameter D_1, the process being viewed as steady state once material begins exiting from the die. Due to incompressibility, a volume of metal $A_o \Delta \ell$ must exit as $A_1 \Delta \ell_1$ and the total actual work, W_a, done externally must be

$$W_a = F_e \Delta \ell \tag{6-5}$$

where F_e is the force applied to the billet. Recognizing that the work per unit volume $w_a = W_a / A_o \Delta \ell$,

$$w_a = \frac{F_e}{A_o} \frac{\Delta \ell}{\Delta \ell}$$

or, defining F_e / A_o as an extrusion pressure, P_e,

$$w_a = P_e \tag{6-6}$$

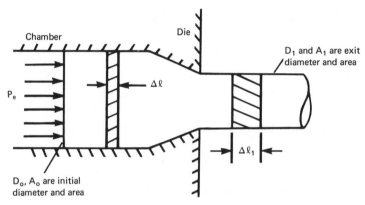

Figure 6-1 Illustration of direct or forward extrusion assuming ideal deformation.

For an ideal process, $w_a = w_i$ and recognizing that in reality $w_a \geq w_i$, the inequality

$$P_e \geq \int \bar{\sigma} \, d\bar{\epsilon} \tag{6-7}$$

results. Setting P_e equal to the integral provides an underestimate of the pressure, that is, the ideal work method gives a *lower bound* of applied forces, pressures, etc.

A similar approach can be followed for rod or wire drawing, as illustrated in Fig. 6-2. Application of a drawing force, F_d, causes metal again to flow through the die, and as that force moves through a distance $\Delta \ell$, the actual work per unit volume is identical in form to Eq. (6-6), that is,

$$w_a = \frac{F_d}{A_1} = \sigma_d \tag{6-8}$$

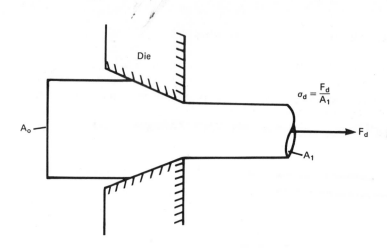

Figure 6-2 Illustration of rod or wire drawing.

where σ_d is the drawing stress. Again noting that $w_a \geq w_i$,

$$\sigma_d \geq \int \bar{\sigma} d\bar{\epsilon} \qquad (6\text{-}9)$$

Example 6-2

If the bar in Ex. 6-1 were drawn through a die to reduce the 12.7-mm diameter to 11.5 mm, determine the drawing stress, σ_d, assuming an ideal process.

Solution. Equation (6-9) is used here where, for an ideal process,

$$\sigma_d = \int_o^\epsilon \bar{\sigma} d\bar{\epsilon}$$

and since all calculations are identical to Ex. 6-1, $\sigma_d = 2650$ psi.

6-4 FRICTION, REDUNDANT WORK, AND MECHANICAL EFFICIENCY

Besides the work required for homogeneous deformation, the actual work needed to produce a shape change also includes those effects of friction and redundant deformation. Frictional work per unit volume, w_f, is consumed at the interface between the deforming metal and the tool faces that constrain the metal. Redundant work, w_r, is due to internal distortion in excess of that needed to produce the desired shape. Figure 6-3 illustrates redundant deformation.

If the deformation were ideal, plane sections would remain plane as the process occurs. In reality, internal shearing causes distortion of plane sections as they pass through the deformation zone. As a consequence, the metal experiences a strain greater than that based upon the ideal shape change; this causes the end product to be harder, stronger, and less ductile than it would be if the

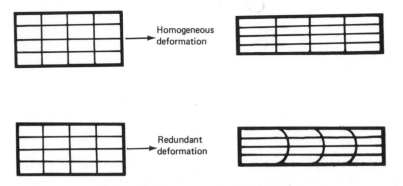

Figure 6-3 Comparison of ideal and actual deformation to illustrate the meaning of redundant deformation.

deformation were truly homogeneous. The redundant work is the energy expended per unit volume to cause this redundant strain.

The net result is that the actual work is the sum of three terms,

$$w_a = w_i + w_f + w_r \tag{6-10}$$

In general, it is difficult in practice to separate w_f from w_r since they are not mutually independent. This can be circumvented by defining a deformation efficiency, η, where

$$\eta \equiv \frac{w_i}{w_a} \tag{6-11}$$

With this definition, the non-ideal work terms are lumped together, and although η is certainly influenced by such factors as the die angle, reduction per pass, and lubrication at the tool-workpiece interface, reasonable estimates for η have been based upon experience. Often in practice, η varies between 0.5 and 0.65. Using an efficiency factor as described, Eqs. (6-7) and (6-9) could be expressed as,

$$P_e \text{ (or } \sigma_d\text{)} = \frac{1}{\eta} \int \bar{\sigma} d\bar{\epsilon} = \frac{w_i}{\eta} \tag{6-12}$$

For power-law hardening, w_i is given by Eq. (6-2). If the effect of work hardening is small, such as in hot working or in cases where the metal to be deformed has already been heavily cold worked, an average value for $\bar{\sigma}$ can be used in Eq. (6-12), such that

$$\frac{1}{\eta} \int_{\epsilon_1}^{\epsilon_2} \bar{\sigma} d\bar{\epsilon} = \frac{\sigma_a(\Delta\epsilon)}{\eta} \tag{6-13}$$

σ_a being the average flow stress over the range $\Delta\epsilon$.

Example 6-3

If the bar in Ex. 6-1 is reduced by extruding and the deformation efficiency is 70%, what is the extrusion pressure and the average flow stress for the operation?

Solution.   From Eq. (6-12),

$$P_e = \frac{1}{\eta} \int_0^{\bar{\epsilon}} \bar{\sigma} d\bar{\epsilon} = \frac{w_i}{\eta} = \frac{2650}{0.7} = 3790 \text{ psi}$$

With Eq. (6-13),

$$\sigma_a = \frac{\eta P_e}{\Delta \epsilon} = \frac{0.7(3790)}{0.199} = 13,350 \text{ psi}$$

6-5 MAXIMUM DRAWING REDUCTION

One use of Eq. (6-12) is to estimate the largest possible reduction that could be made in a single pass during wire or rod drawing. If σ_d needed to effect the drawing operation is greater than the flow stress of the metal at the exit end, that material will yield in tension before flow occurs in the die zone, and if σ_d *exceeds* the tensile strength, drawing is not possible under the existing conditions of expected reduction and die angle.

Two possible conditions, initial start-up and steady-state drawing, are of concern. First consider the steady-state situation where material exiting from the die has been work hardened by its passage through the deformation zone. Often, the strain induced by the drawing operation exceeds the instability strain found in a tension test, so the tensile strength and flow stress (i.e., yield strength) of the drawn material are equivalent. Thus, any yielding would cause immediate necking and failure, and the maximum possible strain due to drawing, designated as ϵ^*, would be reached when $\sigma_d = \bar{\sigma}$. If any added strain due to redundant deformation is neglected for the first drawing pass, it follows from Eq. (6-12) that

$$\sigma_d = \frac{1}{\eta} \int_0^{\epsilon^*} \bar{\sigma} d\bar{\epsilon} \tag{6-14}$$

and for power-law hardening,

$$\epsilon^* = \eta(1 + n) \tag{6-15}$$

For an ideal plastic ($n = 0$) and perfect efficiency ($\eta = 1$), $\epsilon^* = 1$ and the maximum reduction is $r^* = 0.63$ or a 63% area reduction. With a more reasonable value of $\eta = 0.65$, $r^* = 0.48$. In the presence of work hardening ($n > 0$), the maximum reduction is raised by a factor of $(1 + n)$ on the first pass, so theoretically at least, reductions greater than 63% are possible. Practical wire-drawing operations involve successive reductions, so the influence of work hardening decreases from die to die and, as a consequence, a practical limit for ϵ^* under such conditions approaches η since $n \rightarrow 0$.

Regardless of the work-hardening behavior of a metal (that is, it may not be of a power-law type), the drawing limit strain, ϵ^*, is determined by the intersection of the work-hardening curve and the drawing-stress behavior as a function of strain, as illustrated by Fig. 6-4. Drawing-stress behavior is depicted by curves where the efficiency is ideal and when $\eta < 1.0$.

Now consider the situation related to initial start-up. Before the drawing process can begin, a section of undrawn material must be reduced in diameter

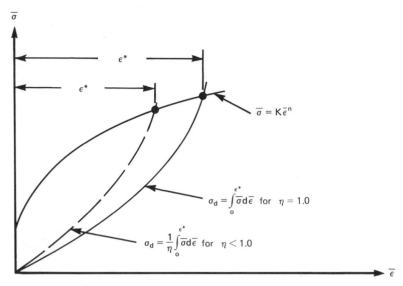

Figure 6-4 The tensile stress-strain curve and the drawing stress-strain behavior for two levels of deformation efficiency. The intersection points, ϵ^*, are the limit strains in drawing.

to fit through the die opening so as to provide an "outlet" section through which the external drawing force can be applied. If this outlet section is produced by machining or etching, the maximum value of σ_d permitted will equal the tensile strength of the outlet material, and the maximum possible reduction, ϵ^*, can be found using Eq. (6-14). If, as is commonly done industrially, the reduced section is developed by rotary swaging, the initial outlet section may be work hardened to such an extent that it exceeds the degree of work hardening induced in the drawn metal. In such a case, the drawing limit is controlled by the same factors discussed for steady-state conditions.

The reductions employed in typical industrial operations are much lower than the maximum possible reductions discussed here, since failure at any one die requires a refeeding of that and all subsequent dies. This is a time consuming and costly operation and is to be avoided if at all possible.

The effects of the half die angle, α, on the actual work, w_a, are presented qualitatively in Fig. 6-5 where the homogeneous, frictional, and redundant work terms are considered separately. For a given reduction, r, the length of contact between the die and the deforming metal increases as α decreases. Since the compressive *stress* between die and workpiece is essentially unchanged, the *force* along that interface must increase with decreasing α, since the *contact area* increases. If the coefficient of friction, μ, is relatively constant, then the frictional work, w_f, increases as α decreases. Regarding the redundant work, w_r, as α increases for a given r, the degree of redundant deformation goes up, so w_r increases accordingly. The ideal work, w_i, being a function of r only, is unaffected by α. Due to the countering effects of α on w_f and w_r, the plot of actual work, w_a, is described by the behavior shown on Fig. 6-5 where some

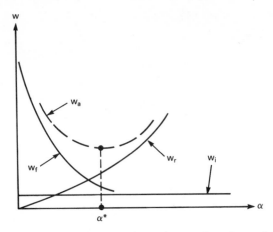

Figure 6-5 Influence of semi-die angle on the actual work, w_a, during drawing where the individual contributions of ideal, w_i, frictional, w_f, and redundant work, w_r, are shown.

optimum angle, α^*, requires the minimum for w_a. Note that as heavier reductions and increased friction are encountered, the actual value for α^* increases. Figures 6-6 and 6-7, adapted from Wistreich's[†] work, illustrate this behavior.

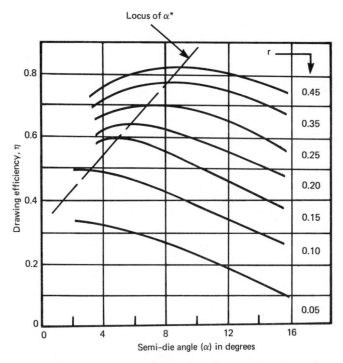

Figure 6-6 Effect of semi-die angle on drawing efficiency for various reductions; note the change in the optimal die angle, α^*.

†J. Wistreich, *Metals Rev.*, 3 (1958), pp. 97–142.

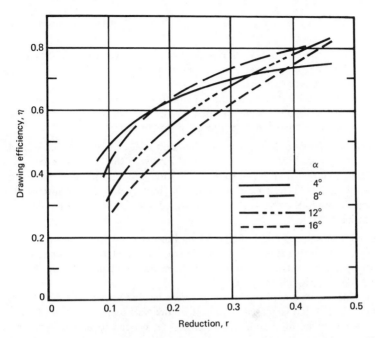

Figure 6-7 Replot of data in Fig. 6-6 showing the effect of reduction on efficiency for various semi-die angles used in drawing.

Compared with wire or rod drawing, extrusion operations have no "theoretical" limiting reductions. In practice, machine capacity and tool strength do impose limits but, in general, reductions by extrusion are much larger than those attempted by drawing, and comparable values of α are therefore also larger.

PROBLEMS

6-1. A round rod of initial diameter D_o can be reduced to diameter D_e by pulling through a conical die with a necessary load, F_d, as shown in sketch (a). A similar result can occur by applying a uniaxial tensile load, as shown in sketch (b).

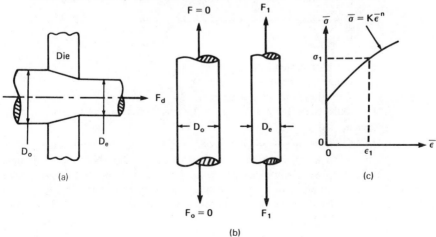

The stress-strain behavior is depicted in (c), the metal obeying $\bar{\sigma} = K\bar{\epsilon}^n$, and σ_1 is to be considered as the true stress needed to reduce D_o to D_e (ϵ_1 is the corresponding true strain).

Using the ideal-work method for both the drawing and tensile operations, compare the load F_d with the load F_1 (or the "drawing stress," σ_d with the tensile stress σ_1) needed to produce equivalent reductions.

6-2. There is a limit to the maximum reduction that can be made in a single wire-drawing pass. Greater reduction would require that the force on the drawn section be larger than it can support. Calculate the maximum reduction possible for a material whose stress-strain curve is given by $\bar{\sigma} = K\bar{\epsilon}^n$, assuming "homogeneous" flow in the die. If $K = 200,000$ and $n = 0.5$, what value of r should be obtainable?

6-3. A billet of an aluminum alloy is being hot extruded from a 4-in. diameter to a 1-in. diameter in a single stroke, as shown in the sketch. If the yield stress of the metal remains constant at 10 ksi (i.e., no work hardening) during the operation and the process efficiency, η, is 0.50,

(a) What is the magnitude of the *pressure* needed to perform the operation?

(b) Calculate the lateral *pressure* felt by the wall of the container (see sketch).

(c) What is the minimum wall thickness, t, needed to prevent yielding of the container walls if the container is made of a metal whose yield strength is 100 ksi?

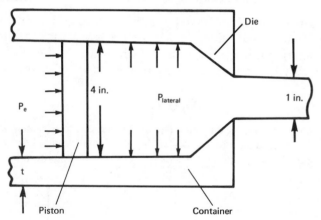

6-4. Regarding the plot shown in Fig. 6-4, $\eta = 1.0$

(a) What important factors are neglected?

(b) How would these factors influence r_{max}?

(c) How would these factors influence the mechanical properties of the final product?

6-5. A method for forming a tapered end on a solid cylindrical bar is suggested in the sketch; it consists of forcing the bar into a conical die whose included angle is 30°. The bar is short enough that buckling may be ignored, but the unformed portion must remain with a 1-in. diameter. If the process efficiency, η, is 65% and strain hardening is ignored, determine the smallest diameter, D_1, that could be achieved.

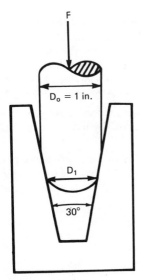

6-6. A sheet 36 in. wide and 0.250 in. thick is to be cold rolled to a thickness of 0.200 in. in a single pass; the 36-in. dimension does not increase during this operation. The work-hardening relation $\bar{\sigma} = 100,000\,\bar{\epsilon}^{0.20}$ pertains for this metal and the inlet speed of the sheet is 1000 ft per minute. A deformation efficiency, η, of 75% is reasonable for this operation and the von Mises yield criterion is applicable. If effects of strain rate are ignored, determine the horsepower required to perform this operation using the ideal-work method of analysis.

6-7. A metal, whose strain-hardening function is given by $\bar{\sigma} = 50,000\,\bar{\epsilon}^{0.3}$, is subjected to a plastic deformation that induces plastic strains of $\epsilon_1 = 0.200$ and $\epsilon_2 = -0.125$. Determine the plastic work per unit volume that is expended during this operation, if the ideal-work method is applicable.

6-8. Your company is planning a wire-drawing operation and you have been asked to determine how many drawing passes are required to reduce copper wire from 0.025 in. to 0.010 in. in diameter.

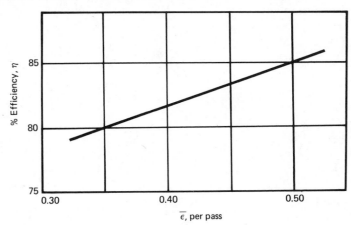

The dependence of deformation efficiency, η, on reduction per pass has been determined from laboratory experiments using the same die angle and lubrication that will be used in production (see sketch). To be sure that no drawing failures occur, assume that the efficiency in production is 75% of that measured in the laboratory, and design the reduction schedule such that the drawing stress never exceeds 60% of the flow stress. Neglect work hardening.

6-9. Using Eq. (6-14), derive an expression for ϵ^* at the initiation of drawing when the outlet section is produced by machining an annealed rod to fit through the die.

6-10. For a certain material, the true stress-strain curve is approximated by $\sigma = B + C\epsilon$, where B and C are constants. In wire drawing, what is the maximum strain that can be made per pass if $B = 30{,}000$ psi, $C = 10{,}000$ psi, and the drawing efficiency is 75%?

$$7$$

Slab analysis—
force balance

7-1 INTRODUCTION

This method, which is also called the free body-equilibrium approach, entails a force balance on a slab of metal of differential thickness. This produces a differential equation where variations are considered in one direction only. Using pertinent boundary conditions, an integration of this equation then provides a solution. The pertinent assumptions involved are:

1. The direction of the applied load and planes perpendicular to this direction define principal directions, and the principal stresses do not vary on these planes.
2. Although effects of surface friction are included in the force balance, these do not influence the internal distortion of the metal or the orientation of principal directions.
3. Plane sections remain plane, thus the deformation is homogeneous in regard to the determination of induced strain.

7-2 SHEET DRAWING

To illustrate this approach consider a sheet of metal having initial width, w, and thickness, t_o, being pulled through a pair of wedges; each wedge is equally inclined to the centerline and $w \gg t_o$. This process is usually called strip or sheet drawing and the details are shown in Fig. 7-1.

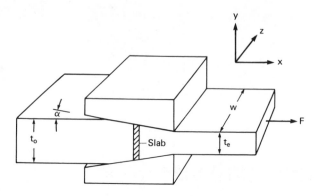

Figure 7-1 Essentials of plane-strain strip or sheet drawing.

Due to the initial geometry of the sheet, the deformation is, for all practical purposes, in plane strain with no change in the width. In essence, the change to the exit thickness, t_e, is accommodated by an equivalent increase in the length of the workpiece. The applied force, F, divided by the exit area, wt_e, is the "drawing" stress, σ_d, i.e., the "exit" stress acting in the x direction. For clarity, a slab of thickness dx is enlarged in Fig. 7-2a.

The virtual apex of the wedges provides an origin with the slab which is located at distance x from this origin. As the stress does not vary on planes perpendicular to the flow direction, any change in σ_x associated with dx is one dimensional as shown. Die pressure normal to the workpiece is given by P, and friction effects at each surface by μP, where μ is the coefficient of friction. In this example, there is no axial backload (or stress) acting opposite to F, but because of the friction and the horizontal component of P, the stress increases from inlet to exit.

Considering the equilibrium of *forces* in the x direction:

$$(\sigma_x + d\sigma_x)(t + dt)w + 2P \sin \alpha \frac{dx}{\cos \alpha} w + 2\mu P \cos \alpha \frac{dx}{\cos \alpha} w = \sigma_x wt \qquad (7\text{-}1)$$

Simplifying, and neglecting differentials of higher order, one obtains

$$\sigma_x dt + t d\sigma_x + 2p dx \tan \alpha + 2\mu P dx = 0 \qquad (7\text{-}2)$$

Substituting $2dx = dt/\tan \alpha$ (see Fig. 7-2), Eq. (7-2) can be expressed as

$$Pdt + \sigma_x dt + t d\sigma_x + \mu P dt \cot \alpha = 0 \qquad (7\text{-}3)$$

or

$$t d\sigma_x + [\sigma_x + P(1 + B)]\, dt = 0 \qquad (7\text{-}4)$$

where $B = \mu \cot \alpha$.

Before Eq. (7-4) can be integrated, a functional relation between P and σ_x must be obtained. If this can be expressed as a function of principal stresses, then invoking a yield criterion will provide the necessary relationship of P and σ_x. With the assumption that x and y are principal directions, σ_x and σ_y are principal stresses and do not vary in the plane of the slab. Taking $\Sigma F_y = 0$ at the inter-

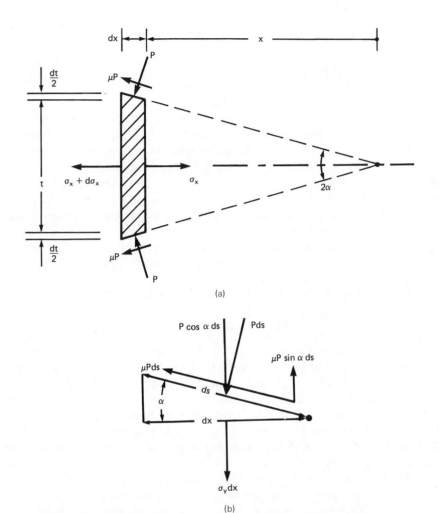

Figure 7-2 Enlarged view of the slab used for a force balance analysis during plane-strain sheet drawing.

face, as in Fig. 7-2b,

$$\sigma_y dx + P \cos \alpha \, ds = \mu P \sin \alpha \, ds \qquad (7\text{-}5)$$

and since $\cos \alpha = dx/ds$, $\sigma_y = -P + \mu P \tan \alpha = -P(1 - \mu \tan \alpha)$. In general, $\mu \tan \alpha \ll 1$, so

$$\sigma_y \approx -P \qquad (7\text{-}6)$$

indicating that σ_y is compressive. In terms of principal stresses, since σ_x is obviously tensile, then

$$\sigma_x = \sigma_1, \; \sigma_y = \sigma_3 = -P, \; \sigma_z = \sigma_2$$

and for plane strain, both the von Mises and Tresca criteria give

$$\sigma_1 - \sigma_3 = 2k = \sigma_x - (-P)$$

where the shear yield stress, k, is $Y/2$ for Tresca and $Y/\sqrt{3}$ for von Mises. Thus,

$$P = 2k - \sigma_x \tag{7-7}$$

Before solving Eq. (7-4) with the aid of Eq. (7-7), several physical points should be noted. It has been assumed that x and y are principal directions; however, if the friction at the surface is significant or if α is large, there must be a shear stress at the surface, so σ_x and σ_y cannot truly be principal stresses. In effect, there would be a rotation of principal axes away from the x-y system; naturally this complicates matters seriously. However, for low values of μ and α, such added complication would be more of an exercise in mathematics than is worthwhile, and there is no need here to be concerned with these subtleties. This does illustrate one of the shortcomings of assuming that plane sections remain plane (they actually curve as the metal proceeds through the deformation zone).

Substituting Eq. (7-7) into (7-4), and rearranging terms,

$$\frac{d\sigma_x}{B\sigma_x - 2k(1 + B)} = \frac{dt}{t} \tag{7-8}$$

There are three parameters in Eq. (7-8) that could depend upon the position of x. Certainly $2k$ is not constant if work hardening is considered and, if curved dies were used, the angle α would vary with position. Finally, μ can also vary with position at the interface. If any dependence of these variables with x were accurately described, they could be introduced into Eq. (7-8). However, the real functional dependence of μ is generally uncertain at best and the proper form of strain hardening might be questionable. A sensible engineering question at this point would be "Why complicate an admittedly approximate model of real behavior with additional uncertainties?" After all, if one must again *guess* at the correct functional dependence of μ and $2k$ on position, then added mathematical elegance to a problem whose initial formulation rests upon simplifying but not completely realistic assumptions is just not worth the effort.

With these remarks, the solution of Eq. (7-8) is based upon the following:

1. A gross, average constant value of μ describes the full contact region.
2. The metal does not work harden, or a "mean" value of yield shear strength adequately describes any work-hardening effects; in either case, $2k$ is treated as a constant.
3. The die angle α is a constant.

Direct integration of Eq. (7-8). using the conditions that $\sigma_x = 0$ when $t = t_o$ and $\sigma_x = \sigma_d$ when $t = t_e$, gives

$$\sigma_d = \frac{2k(1+B)}{B}\left[1 - \left(\frac{t_e}{t_o}\right)^B\right] \tag{7-9}$$

Since the homogeneous strain, $\epsilon_h = \ln(t_o/t_e)$, another form of Eq. (7-9) is

$$\frac{\sigma_d}{2k} = \frac{1+B}{B}[1 - \exp(-B\epsilon_h)] \tag{7-10}$$

Note that $\epsilon_h = -\epsilon_y = \epsilon_x$. In fact, Eq. (7-10) can be obtained initially if in Eq. (7-8) dt/t is expressed in terms of $d\epsilon_h$. In Eq. (7-10), the left-hand side is expressed as a dimensionless stress ratio, which is helpful as a parameter when plotting experimental results.

Example 7-1

A sheet of metal having an initial thickness of 0.100 in. and width of 12 in. is to be drawn through straight-sided dies having an included angle of 30°. If the average of the yield shear stress is 30 ksi and an average value for the coefficient of friction is 0.08, calculate the force needed to complete this operation for a reduction of 10%.

Solution.   Using Eq. (7-10) and noting that $B = \mu \cot \alpha$, where α is the semi-die angle,

$$B = 0.08 \cot 15° \approx 0.3$$

$$\epsilon_h = \ln\left(\frac{1}{1-r}\right) = \ln\left(\frac{1}{1-0.1}\right) = 0.105$$

From Eq. (7-10),

$$\sigma_d = 2(30)\left(\frac{1+0.3}{0.3}\right)[1 - \exp(-0.3 \times 0.105)]$$

$$\sigma_d = 8.08 \text{ ksi}$$

The drawing force, $F_d = \sigma_d(w)t_e$, so

$$F_d = 8,080(12)(0.09) = 8,730 \text{ lbf}$$

7-3 COMPARISON OF SLAB METHOD AND IDEAL-WORK METHOD

Although the mathematics are left as an exercise, the problem just discussed provides a convenient illustration for comparing the predictions by using the two techniques covered to this point. If μ is taken as zero in Eq. (7-4) or if L'Hospital's rule is applied to Eq. (7-10) where $B = 0$, there results

$$\frac{\sigma_d}{2k} = \ln\left(\frac{t_o}{t_e}\right) = \epsilon_h \tag{7-11}$$

Using the ideal-work method provides exactly the same result; thus, if friction is neglected, either method leads to an identical finding. From this it can be realized that the force-balance method simply extends the information provided by the ideal-work method to include possible frictional effects.

7-4 WIRE OR ROD DRAWING

This analysis was first developed by Sachs[†] and parallels the development just shown which led to Eq. (7-10). For a wire or rod of circular cross section, the basic governing differential equation is:

$$\frac{d\sigma}{B\sigma - (1 + B)\bar{\sigma}} = 2\frac{dD}{D} \tag{7-12}$$

Note the similarity with Eq. (7-8). Now

$$2\frac{dD}{D} = -d\epsilon$$

so that

$$\int_o^{\sigma_d} \frac{d\sigma}{B\sigma - (1 + B)\bar{\sigma}} = -\int_o^{\epsilon_h} d\epsilon \tag{7-13}$$

Integrating gives

$$\sigma_d = \sigma_a\left(\frac{1 + B}{B}\right)[1 - \exp{(-B\epsilon_h)}] \tag{7-14}$$

where σ_a may be taken as the mean yield stress, Y_m, over the range of strain, ϵ_h, induced by the shape change.

It should be noted that these analyses become unrealistic at high die angles and low reductions. Assuming that P is a principal stress is reasonable only if α is small and friction is low. Additionally, neglecting redundant strain is also questionable for combinations of large die angles and small reductions.

7-5 DIRECT COMPRESSION IN PLANE STRAIN

As another illustration of the slab analysis, consider the compression of a block under conditions of plane-strain deformation. Figure 7-3 illustrates the operation, where $h \ll b$ and sliding friction with a constant coefficient μ is assumed to prevail at the interfaces. Considering a force balance on the slab in the x direction,

$$\sigma_x h + 2\mu P dx - (\sigma_x + d\sigma_x)h = 0 \tag{7-15}$$

$$2\mu P dx = h d\sigma_x \tag{7-16}$$

Again, σ_x and σ_y (taken as $-P$) are considered as principal stresses. Note that with low μ, the shear influence is again considered to be small in regard to principal stress directions. For plane strain, $\sigma_x - \sigma_y = 2k$ or $\sigma_x + P = 2k$, therefore, $d\sigma_x = -dP$. Now Eq. (7-16) can be expressed as

†G. Sachs, *Z. agnew. Math. Mech.*, 7, p. 235 (1927).

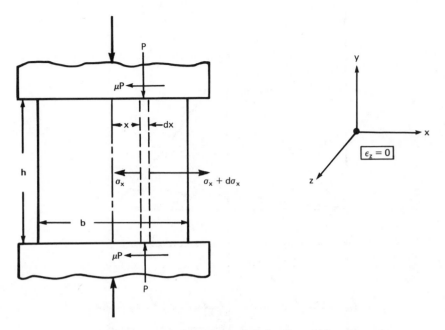

Fig. 7-3 Essentials for a slab force analysis in plane-strain compression.

$$2\mu P dx = -h dP \qquad (7\text{-}17)$$

or
$$\frac{dP}{P} = \frac{-2\mu}{h} dx$$

where at $x = b/2$, $\sigma_x = 0$ and $P = 2k$, so the solution is

$$\ln\left(\frac{P}{2k}\right) = \frac{2\mu}{h}\left(\frac{b}{2} - x\right) \qquad (7\text{-}18)$$

or
$$\frac{P}{2k} = \exp\left[\frac{2\mu}{h}\left(\frac{b}{2} - x\right)\right] \qquad (7\text{-}19)$$

which governs from $x = 0$ to $x = b/2$. The maximum value for P occurs at the centerline, where

$$\left(\frac{P}{2k}\right)_{\text{max}} = \exp\left(\frac{\mu b}{h}\right) \qquad (7\text{-}20)$$

A plot of $P/2k$ versus x is shown in Fig. 7-4, where the rise in P towards the centerline is referred to as the *friction hill*.

Of great interest is the mean or average pressure at the tool-workpiece interface. This pressure, P_a, times the contact area leads to a calculation of the actual force that must be applied.

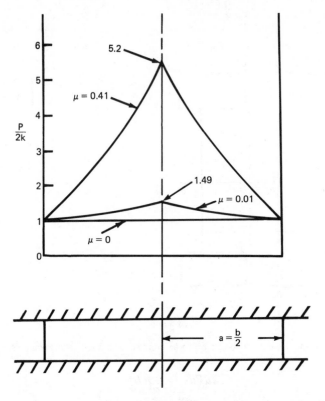

Figure 7-4 Illustration of the friction hill in plane-strain compression for different values of the coefficient of friction.

7-6 AVERAGE PRESSURE DURING PLANE-STRAIN COMPRESSION

For simplicity, let $a = b/2$ and $c = 2\mu/h$ in the following derivation, and note that e is the base of natural logarithms.

$$P_a = \frac{1}{a} \int_o^a P\,dx = \frac{1}{a} \int_o^a 2ke^{c(a-x)}dx = \frac{2ke^{ca}}{a} \int_o^a e^{-cx}dx$$

so

$$P_a = \frac{2ke^{ca}}{a}\left[\frac{-e^{-cx}}{c}\right]_{x=o}^{x=a}$$

Thus,

$$P_a = \frac{2k}{\mu b/h}(e^{\mu b/h} - 1) \tag{7-21}$$

Now,

$$\exp\left(\frac{\mu b}{h}\right) - 1 = 1 + \frac{\mu b}{h} + \frac{(\mu b/h)^2}{2} + \ldots - 1$$

so

$$P_a \approx 2k\left(1 + \frac{\mu b/h}{2} + \ldots\right) \approx 2k\left(1 + \frac{1}{2}\frac{\mu b}{h}\right) \text{ for small } \frac{\mu b}{h} \tag{7-22}$$

It is important to note an upper limit for the term μP in this development. To avoid shearing of the workpiece at the interface, $\mu P \leq k$, which is the yield strength in shear. As indicated in Fig. 7-4, $P \geq 2k$, thus, $\mu \leq \frac{1}{2}$ if sliding friction is to occur.

Example 7-2

Plane-strain compression is conducted on a slab of metal whose yield shear strength, k, is 15,000 psi. The width of the slab is 8 in. while its height is 1 in. Assuming the average coefficient of friction at each interface is 0.10,
a) Estimate the maximum pressure at the onset of plastic flow, and
b) Estimate the average pressure at the onset of plastic flow.

Solution
a) From Eq. (7-20),

$$P_{\max} = 2k \exp \left(\frac{\mu b}{h}\right) = 30,000 \exp \left(\frac{0.1 \times 8}{1}\right)$$

so, $P_{\max} = 30,000(2.226) = 66,800$ psi.
b) First, use the exact solution given by Eq. (7-21):

$$P_a = \frac{30,000}{0.8}(e^{0.8} - 1) = 46,000 \text{ psi}$$

Next use the approximation given by Eq. (7-22):

$$P_a = 30,000(1 + 0.4) = 42,000 \text{ psi}$$

This indicates that a value of 0.8 for $(\mu b)/h$ should not be considered small in this context. For instance, if b were 4 in., h were 1.5 in., and μ were 0.05, then $(\mu b)/h$ would be 0.133. The respective values of P_a using Eqs. (7-21) and (7-22) would then be 32,100 and 32,000 psi, which are nearly the same.

7-7 STICKING FRICTION AT THE INTERFACE

As just indicated, there is a limit at which sliding friction can exist at the tool-workpiece interface: if that is reached, then interfacial shear of the workpiece occurs and the frictional forces indicated as μP in Fig. 7-3 are replaced by the yield shear stress, k, of the workpiece in Eq. (7-16). Following the analysis that produced Eq. (7-19), the result for sticking friction is

$$\frac{P}{2k} = 1 + \frac{b/2 - x}{h} \tag{7-23}$$

which predicts a linear variation of P from the outer edge to the centerline as demonstrated in Fig. 7-5; the maximum value, which occurs at the centerline, is

$$P_{\max} = 2k \left(1 + \frac{b}{2h}\right) \tag{7-24}$$

If in Fig. 7-5, $b = 4h$, then $P_{\max} = 6k$ and $P_{\text{avg}} = 4k$ as shown.

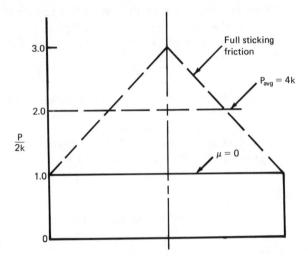

Figure 7-5 The friction hill in plane-strain compression for sticking friction.

For the general case, and following procedures as in Sec. 7-6, the average pressure with sticking friction can be shown to be

$$P_a = 2k\left(1 + \frac{b}{4h}\right) \qquad (7\text{-}25)$$

Example 7-3
Repeat Ex. 7-2 if sticking friction prevailed at each interface.

Solution
a) Using Eq. (7-24),

$$P_{\max} = 30,000\left(1 + \frac{8}{2}\right) = 150,000 \text{ psi}$$

b) Using Eq. (7-25),

$$P_a = 30,000\left(1 + \frac{8}{4}\right) = 90,000 \text{ psi}$$

This can also be seen from Fig. 7-5, where in this problem $(P/2k)_{\max}$ equals 5. So $P_a/2k = [\frac{1}{2}(\text{base})(5 - 1)]/\text{base} + 1 = 3$, or $P_a = 6k = 90,000$ psi.

7-8 CONSTANT INTERFACIAL SHEAR STRESS

When lubrication is achieved by placing a film of a soft solid (e.g., a polymer) between the tools and workpiece, the frictional conditions are better approximated by assuming a constant shear stress, $\tau = mk$, at the interface where $0 < m < 1$. Then Eqs. (7-23), (7-24), and (7-25) would be identical in this instance except that m would be a coefficient of each term containing b.

7-9 AXISYMMETRIC COMPRESSION

Referring to Fig. 7-6 and using an approach similar to that in Sec. 7-5, the element to be analyzed is a ring of differential thickness, dr, and height, h, at some radius, r, from the center.

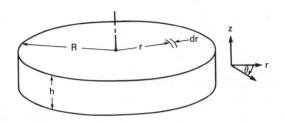

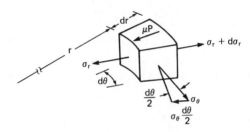

Figure 7-6 Element used in a slab analysis for axisymmetric compression.

For a constant value of μ, a force balance in the radial direction gives

$$\sigma_r h r d\theta + 2\mu P r d\theta dr + \frac{2\sigma_\theta h dr d\theta}{2} - (\sigma_r + d\sigma_r)hd\theta(r + dr) = 0 \qquad (7\text{-}26)$$

or

$$2\mu P r dr + h\sigma_\theta dr - h\sigma_r dr - h r d\sigma_r = 0 \qquad (7\text{-}27)$$

if higher order terms are neglected.

With axisymmetric flow, $\epsilon_\theta = \epsilon_r$ so $\sigma_\theta = \sigma_r$, and for yielding,[†] $\sigma_r + P = Y$ or $d\sigma_r = -dP$. Inserting these into Eq. (7-27) gives

$$2\mu P r dr = -h r dP$$

or

$$\frac{dP}{P} = -\frac{2\mu}{h} dr \qquad (7\text{-}28)$$

Since $P = Y$ when $r = R$, the solution of Eq. (7-28) is

$$P = Y \exp\left[\frac{2\mu}{h}(R - r)\right] \qquad (7\text{-}29)$$

The average pressure is

[†]Using the Tresca criterion, $Y = 2k$, whereas with the von Mises criterion, $Y = \sqrt{3}k$.

$$P_a = \frac{1}{\pi R^2} \int_0^R P \cdot 2\pi r dr = \frac{2Y}{R^2} \int_0^R r \exp\left[\frac{2\mu}{h}(R-r)\right] dr \qquad (7\text{-}30)$$

using Eq. (7-29). The solution of Eq. (7-30) is

$$P_a = \frac{1}{2}\left(\frac{h}{\mu R}\right)^2 Y\left[\exp\left(\frac{2\mu R}{h}\right) - \frac{2\mu R}{h} - 1\right] \qquad (7\text{-}31)$$

For small values of μ and moderate values of R/h, a good first approximation of Eq. (7-31) is

$$P_a = Y\left(1 + \frac{a}{3} + \frac{a^2}{12} + \cdots\right) \qquad (7\text{-}32)$$

where $a = 2\mu R/h$ and only the first three terms in the expansion of e^a are used.

If sticking friction occurs at the interface, there would be a constant shear stress, k. The friction term in Eq. (7-27) is $2krdr$, and Eq. (7-28) becomes

$$2kdr = -hdP \qquad (7\text{-}33)$$

so

$$\int_r^R 2kdr = -\int_P^Y hdP$$

or

$$P = Y + \frac{2k}{h}(R-r) \qquad (7\text{-}34)$$

Following the procedures that led to Eq. (7-31), the average pressure for constant shear stress is

$$P_a = Y + \frac{2kR}{3h} \qquad (7\text{-}35)$$

Example 7-4

A solid disc of 4-in. diameter and 1-in. height is to be compressed as discussed in relation to Fig. 7-6. If the tensile and shear yield stresses for this metal are 50,000 and 25,000 psi respectively, estimate the force needed to start plastic flow.

Solution.   The average pressure at the start of flow is found using Eq. (7-35).

$$P_a = 50,000 + \frac{2(25,000)(2)}{3(1)}$$

so

$$P_a = 50,000 + 33,333 = 83,333 \text{ psi}$$

The force is equal to this pressure times the initial area, so

$$F = 83,333\left(\frac{\pi}{4}\right)(4)^2 \approx 1.05 \times 10^6 \text{ lbf}$$

7-10 SAND-PILE ANALOGY†

The analyses for plane-strain and axisymmetric compression with sticking friction or a constant shear stress interface can be interpreted in terms of a sand-pile analogy. If dry sand is piled upon a flat area, the pile will develop with a constant slope which is characteristic of the type of sand, so that the height

†Many texts call this the sand-heap analogy.

increases linearly with the distance from the nearest edge of the pile. This is analogous to the linear increase of P with distance from the edge of a body being compressed between flat platens. This analogy suggests that the effect of sticking friction, or constant shear stress interface, in the compression of a complex shape can be analyzed by cutting the shape out of cardboard, piling sand upon it, and then measuring the volume of the sand pile and the angle of inclination of the sand. The volume can be determined by pouring the sand into a calibrated beaker, or by weighing if the bulk density is known. The sand volume is equivalent to the integral of $(P - Y)$ over the surface, and hence is a measure of the total compressive force less the compressive force needed in the absence of friction. Shapes of complex cross section can be analyzed in this manner. For a detailed discussion of this technique see Johnson and Mellor [1].

7-11 EXPERIMENTAL FINDINGS

At this point one may ask, "Is it better to assume a constant coefficient of friction or a constant shear stress?" Certainly during most hot-working operations, lubrication is absent or so poor that sticking friction is a reasonable assumption. However, in cold working the answer is not so clear. The little experimental work has indicated that neither assumption is correct.

Measurements of local pressure have been made in axisymmetric compression by imbedding pressure-sensitive pins into the compression platens [2,3]. These measurements (see Fig. 7-7, for example) show that P does increase from

Figure 7-7 Variation of local pressure with radial position for axisymmetric compression using unlubricated and lubricated conditions. These data were abstracted from reference [3] at a 20% reduction.

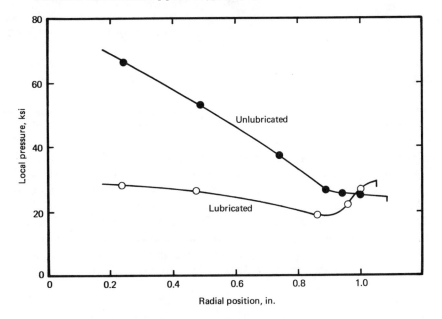

edge to center but that the slope *decreases* toward the center instead of increasing, as a constant coefficient of friction predicts, or remaining constant, as implied by a constant shear stress.

This has been explained as follows: during early compression, lubricant near the edge runs out; the edge of the workpiece then makes contact with the platens and sticks to them. This traps lubricant in the interior, so that the frictional shear stresses toward the center are lower than at the edge. As compression continues and the specimen expands laterally, material which was originally on the vertical sides of the specimen folds onto the compression platens, forming more unlubricated area at the periphery, as shown in Fig. 7-8. Even when a sheet of plastic or soft metal is placed between the workpiece and platens, in an attempt to create a low shear-stress interface, the edges of the workpiece cut through the lubricating sheet at relatively low strains. Thus, neither simple assumption is correct; however, the conclusion that the average compressive pressure, P_a, increases rapidly as a function of (b/h) or (R/h) is still valid.

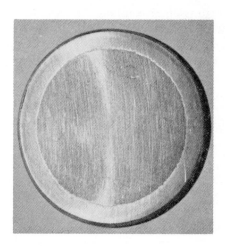

Figure 7-8 Surface appearance of an aluminum disc subjected to compression. The outer annular-like band was originally part of the cylindrical surface and sticking occurred here (from reference [3]).

7-12 FLAT ROLLING

The typical flat rolling of plates, sheets, or strips is essentially a plane-strain operation, since little widening of the workpiece results. This can be explained by examining the shape of the deformation zone as shown in Fig. 7-9. The length of contact, L, between the rolls and workpiece is usually much smaller than the width of the sheet, w. As the rolls induce a compressive stress, σ_z, the plastic zone is thinned and is free to expand in the rolling direction, x, but lateral expansion in the y direction is effectively restrained by the undeforming material on both sides of the roll gap. The net effect is a condition of plane strain, where $\epsilon_y \approx 0$ and $\epsilon_z \approx -\epsilon_x$, except at the edges of the workpiece.

The effects of roll geometry and friction on the rolling process can be understood in terms of the friction-hill effect. Consider the rolling geometry

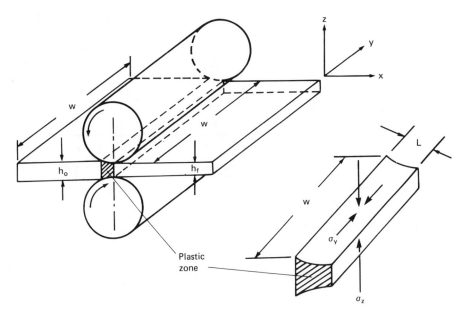

Figure 7-9 Schematic of the deformation zone in flat rolling.

in Fig. 7-10, where at some point, N, in the roll gap, the surface velocities of the rolls and work material are equal. To the left of this point the surface velocity of the metal is lower, and friction between the metal and the rolls tends to draw metal into the gap. To the right of N the metal velocity is higher than the roll velocity, so friction acts to the left on the metal. The net effect is to produce a friction hill.

Consider the roll-gap geometry in Fig. 7-11, where R is the roll radius, $\Delta h = h_o - h_f$ is the decrease in thickness, Q is the projected contact length, and L is the chord of the arc of contact. It is seen that

$$L^2 = Q^2 + \left(\frac{\Delta h}{2}\right)^2$$

and

$$Q^2 = R^2 - \left(R - \frac{\Delta h}{2}\right)^2 = R\Delta h - \left(\frac{\Delta h}{2}\right)^2 \qquad (7\text{-}36)$$

Thus

$$L = \sqrt{R\Delta h} = \sqrt{Rrh_o} \qquad (7\text{-}37)$$

The frictional effects are similar to those in plane-strain compression; in fact, if the roll curvature is neglected, the material in the roll gap can be considered as being under plane-strain compression. With sliding friction, Eq. (7-19) is applicable for the variation in the rolling pressure and Eq. (7-21) for the average roll pressure. For b in those equations, the roll contact length, $L = \sqrt{R\Delta h}$, should be used, and h is the average height of metal in the roll gap (i.e., $h = (h_o + h_f)/2$). Thus in terms of these parameters,

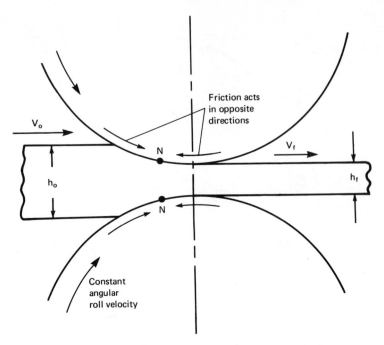

Figure 7-10 Schematic of flat rolling showing the neutral point, N.

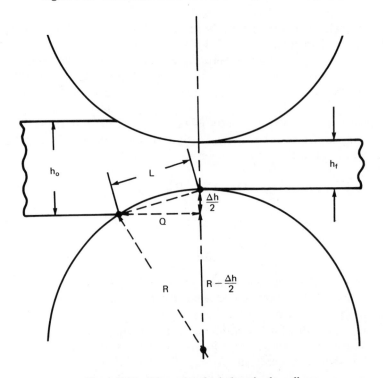

Figure 7-11 Dimensional relations in the roll gap.

$$P_a = \frac{h}{\mu L}\left(\exp\frac{\mu L}{h} - 1\right)\sigma_o \qquad (7\text{-}38)$$

where σ_o is the average plane-strain flow stress $(2k)$ across the roll gap. If the material work hardens, a reasonable and simple approximation of σ_o is $(\sigma_1 + \sigma_2)/2$, where σ_1 and σ_2 are the flow stresses at the entrance and exit of the roll gap. If front tension, σ_{ft}, or back tension, σ_{bt}, is applied during rolling, the compressive stress, P, from the roll is lowered accordingly as shown in Fig. 7-12c. This can be handled approximately by replacing σ_o in Eq. (7-38) by $(\sigma_o - \sigma_t)$, where $\sigma_t = (\sigma_{ft} + \sigma_{bt})/2$. Then

$$P_a = \frac{h}{\mu L}\left(\exp\frac{\mu L}{h} - 1\right)(\sigma_o - \sigma_t) \qquad (7\text{-}39)$$

Example 7-5
The plane-strain flow stress, σ_o, of a metal is 30,000 psi. A sheet of this metal, 24 in. wide by $\frac{1}{8}$ in. thick is to be cold rolled to 0.100 in. in a single pass using rolls of 12-in. diameter, and the coefficient of friction is about 0.075.
a) Compute the average pressure between the rolls and the sheet.
b) If a front tension of 10,000 psi were applied, what would be the average pressure?

Solution
a) For use in Eq. (7-38),

$$h = \frac{0.125 + 0.100}{2} = 0.1125 \text{ in.}$$

$$L = \sqrt{6(0.025)} = 0.387 \text{ in.}$$

$$P_a = \frac{0.1125}{(0.075)(0.387)}\left[\exp\left(\frac{0.075 \times 0.387}{0.1125}\right) - 1\right]30,000$$

$$P_a = 1.14(30,000) = 34,230 \text{ psi}$$

b) From Eq. (7-39) and with the same coefficient from part (a),

$$P_a = 1.14(30,000 - 5,000) = 28,500 \text{ psi}$$

Noting the friction hill on Fig. 7-12c, increasing either front or back tension causes the hill to decrease, and therefore lowers the average pressure. The neutral point is also affected by such tensions. As back tension increases, the shift of this point is towards the exit; lowering friction or increasing reduction has the same tendency towards shifting the neutral point. Regarding an increase in front tension, increasing friction or decreasing reduction all tend to shift the neutral point towards the entry of the material.

If the sheet is thin (small h) and the roll diameter is large (making L large), a large friction hill results and causes a high value of P_a. The roll-separating force, per unit w, $F_s = P_a L$ increases even more rapidly. One consequence of high roll-separating forces is that they cause an elastic flattening of the rolls, much as an automobile tire flattens under the weight of a car. The actual radius of curvature of the roll, R', of the contact area between roll and workpiece

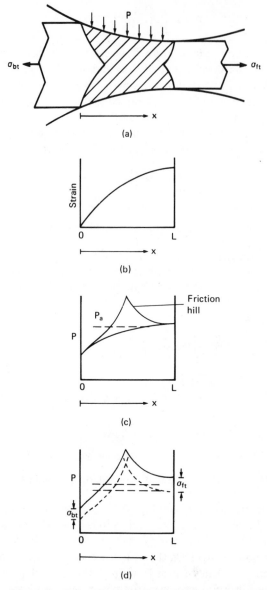

Figure 7-12 Schematic of the roll gap (a) showing the variation of strain (b) and roll pressure (c). The effect of front or back tension upon the friction hill is shown in (d).

becomes larger than the unloaded roll radius, R, as indicated in Fig. 7-13. Hitchcock [4] derived an approximate solution as

$$R' = R\left(1 + \frac{16F_s}{\pi E' \Delta h}\right) \tag{7-40}$$

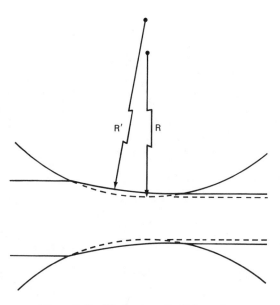

Figure 7-13 Illustration of roll flattening.

where $E' = E/(1 - \nu^2)$. Now using the flattened roll radius, $L \approx \sqrt{R'\Delta h}$ so the separating force per unit length of the roll becomes

$$F_s = P_a(R'\Delta h)^{1/2} \tag{7-41}$$

where

$$P_a = \frac{h}{\mu(R'\Delta h)^{1/2}}\left[\exp\frac{\mu}{h}(R'\Delta h)^{1/2} - 1\right](\sigma_o - \sigma_t) \tag{7-42}$$

so

$$F_s = \frac{h}{\mu}\left[\exp\frac{\mu}{h}(R'\Delta h)^{1/2} - 1\right](\sigma_o - \sigma_t) \tag{7-43}$$

The effect of roll flattening is to increase the separating force, since both P_a and the contact length, L, increase. Both F_s and R' can be found by solving Eqs. (7-40) and (7-43) simultaneously. Figure 7-14 is a graphical solution for heavily rolled steel where $\sigma_o = 100{,}000$ psi, $E' = 33 \times 10^6$ psi, $r = 5\%$ reduction, and $R = 5$ inches. Initial thicknesses (h_o) of 0.100, 0.040, and 0.020 were used in the calculations, and μ was taken as 0.2. Equation (7-40) was used to evaluate R' as a function of F_s whereas Eq. (7-43) gave F_s as a function of R'. The intersection of these two plots satisfies both equations and the operating conditions. There is no intersection point for the sheet where $h_o = 0.020$ inches; this is clarified in the next paragraph.

Roll flattening can become so severe that it is impossible to reduce the thickness of a sheet below some limiting value, $h_{\min}$. The results found by various people give the following:

$$h_{\min} = \frac{C\mu R}{E'}(\sigma_o - \sigma_t) \tag{7-44}$$

with a value of C between 7 and 8.

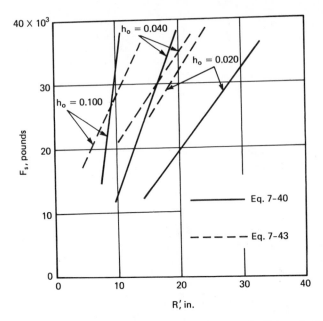

Figure 7-14 Variation of roll force per unit width versus the flattened roll radius for three values of starting sheet thickness. The intersection points satisfy both Eqs. (7-40) and (7-43). Note there is no intersection for $h_o = 0.02$.

Using $C = 7$ and the same conditions cited in the preceding paragraphs, $h_{min} = 0.021$ in.; this explains why there is no intersection point for $h_o = 0.020$ in. on Fig. 7-14. Equation (7-44) suggests certain useful practices when rolling thin sheet or foil. The application of forward and back tension, reducing μ by lubrication, lowering σ_o by annealing and using small-diameter rolls are beneficial in terms of reducing h_{min}. Of course, smaller rolls are subject to greater bending under high separating forces, so back-up rolls are often used to counter this behavior, as exemplified by a Sendzimir mill in Fig. 7-15.

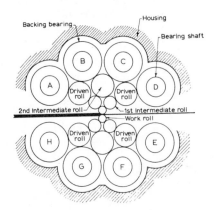

Figure 7-15 Schematic of a Sendzimir mill for rolling thin sheet metal. Courtesy of T. Sendzimir Inc. and the Loewy Robertson Engr. Co. Ltd., U.K.

Another way to reduce elastic roll deflection is to use a material of high elastic modulus, such as sintered carbide, for the work rolls. Still another approach is to place the material to be rolled between layers of softer material in the form of a sandwich, thereby artificially increasing the thickness. Common aluminum foil is rolled as a two-sheet sandwich. The shiny surface results from burnishing by the rolls, the matte surface from contact with the other aluminum sheet.

Example 7-6

Consider a sheet of steel whose $\sigma_o = 100$ ksi, being rolled to a 5% reduction with rolls of 10-in. diameter. The plane-strain modulus, E', is about 33×10^6 psi and the initial sheet thickness is 0.100 in. If the coefficient of friction is 0.15 and R' is about 10 in., find the roll-separating force per unit width of roll:

a) using Eq. (7-40),
b) using Eq. (7-43), and
c) compare the findings with Fig. 7-14.
d) If σ_o were 50 ksi and other conditions were constant, to what minimum thickness could this sheet be rolled?

Solution

a) $$R' = R \left(1 + \frac{16F_s}{\pi E' \Delta h} \right)$$

so $$F_s = \left(\frac{R'}{R} - 1 \right) \frac{\pi E' \Delta h}{16}$$

$$F_s = (2 - 1) \pi \frac{33 \times 10^6 \times 0.005}{16}$$

$$F_s = 32,400 \text{ lbf}$$

b) $$F_s = \frac{0.0975}{0.15} \left\{ \exp \left[\frac{0.15}{0.0975} (10 \times 0.005)^{1/2} \right] - 1 \right\} 100,000$$

$$F_s \approx 27,000 \text{ lbf}$$

c) Note the left-hand pair of lines whose intersection is at an R' of about 9 in. and F_s of about 26,000 lbf. The line with the steeper slope has F_s of about 32,000 when R' is 10 in. as found in part (a) above, while the other line has F_s of about 27,000 when R' is 10 in. as in part (b) above.

d) With Eq. (7-44) and using an average C of 7.5,

$$h_{\min} = \frac{(7.5)(0.15)(5)(50,000)}{33 \times 10^6}$$

$$h_{\min} = 0.0085 \text{ in.}$$

To counter the effects of roll bending, which would produce a sheet of varying thickness across the rolled width, rolls are usually cambered as shown in Fig. 7-16. The degree of camber varies with the width of the sheet, the flow stress of the material, and reduction per pass. Lack of camber or an insufficient amount of camber leads to the result shown in Fig. 7-17a, where the thicker center means that the edges would be plastically elongated more than the center.

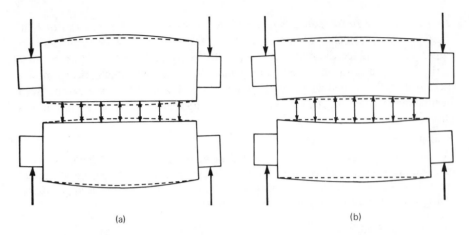

Figure 7-16 (a) Use of cambered rolls to compensate for roll bending. (b) With uncambered rolls, note the variation of thickness.

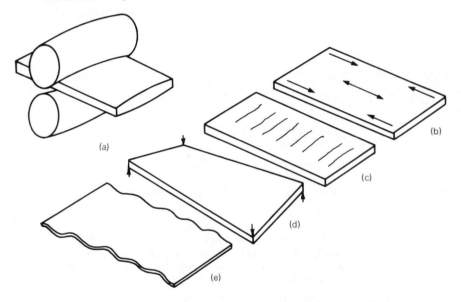

Figure 7-17 Possible effects when rolling with insufficient camber.

This induces a residual stress pattern of compression at the edges and tension along the centerline, shown in Fig. 7-17b. This may cause centerline cracking (Fig. 7-17c), warping (Fig. 7-17d), or edge wrinkling (Fig. 7-17e), sometimes called a crepe-paper effect.

If the rolls are over-cambered, as shown in Fig. 7-18a, the residual stress pattern is opposite that just discussed, i.e., centerline compression and edge tension as in Fig. 7-18b. This may cause edge cracking as in Fig. 7-18c, splitting as in Fig. 7-18d, or centerline wrinkling as in Fig. 7-18e.

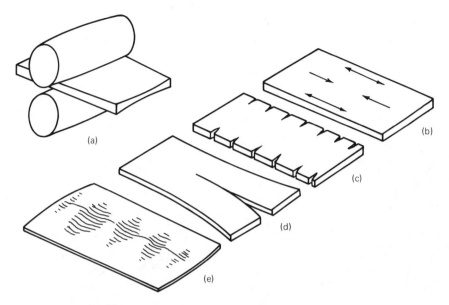

Figure 7-18 Possible effects when rolls are over-cambered.

There are large economic incentives for assuring proper camber in addition to requirements of flatness and freedom from cracks. A thickness variation of only ± 0.001 in. in a 0.032-in. thick sheet between center and edge amounts to a 3 % difference. If a minimum thickness is required, some of the sheet will be thicker than necessary, and either the customer or the supplier (depending on whether the steel is sold by weight or area) will have to pay for several percent more material than actually needed.

Even with proper cambering, there is a tendency for edge cracking caused by the different stress state at the very edge of the sheet, as shown in Fig. 7-19. Except near the edges, undeforming material just outside the plastic zone imposes plane-strain ($\epsilon_y = 0$) deformation in the plastic zone by putting the plastic zone under a lateral (y-direction) compressive stress, $\sigma_y \approx \frac{1}{2}\sigma_z$. However, at the edges σ_y drops to zero, so plane strain does not prevail at the edges. If the edges approach a condition of uniaxial compression, then $\epsilon_y = \epsilon_x$, so the elongation strain at the edge, i.e., ϵ_x, is only one half the value of ϵ_x in the interior. Since, however, the edge and interior must elongate equally, the edge feels a tensile stress, $\sigma_x \approx -\sigma_z$; this tensile effect can cause edge cracking, although the effect is not as great as that induced by over-cambering. Failure to maintain square edges will increase the edge-cracking tendency. With a bulged edge, material at the mid-plane will experience even less compression from the rolls, so increased x-direction tension is necessary to maintain the same elongation as experienced by the rest of the sheet. Bulged edges will, therefore, crack at smaller reductions. In multiple-pass rolling, small edge rolls are commonly used to prevent development of large bulges.

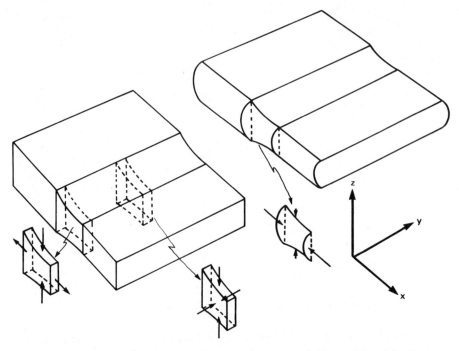

Figure 7-19 Variation of states of stress in the roll gap that may lead to edge cracking.

REFERENCES

[1] W. Johnson and P. B. Mellor, *Engineering Plasticity*. New York: Van Nostrand Reinhold, 1973, pp. 183–90.

[2] G. T. van Rooyen and W. A. Backofen, *Jour. Mech. Phys. Solids*, 7, (1959), pp. 163–68.

[3] G. W. Pearsall and W. A. Backofen, *Trans. ASME 85B* (1963), pp. 68–75.

[4] J. Hitchcock, "Roll Neck Bearings," Appendix I, *Am. Soc. Mech. Eng.*, New York (1935), pp. 286–96.

PROBLEMS

7-1. A coil of steel, 10 in. wide by $\frac{1}{8}$ in. thick is to be drawn through dies of semi-angle 8° to a final thickness of 0.100 in. by a single pass. The outlet speed of the drawn strip is 50 ft per minute, μ is about 0.06, and the average or mean yield strength (i.e., initial plus final divided by 2) is 50,000 psi, which accounts for work hardening and strain-rate effects. If any spreading of the width is negligible, determine the horsepower required to make this reduction. The Tresca yield criterion applies here.

7-2. With a particular metal, rod-drawing experiments indicated an efficiency, η, of 65% when $r = 0.2$ and a semi-die angle of 6° was used.

(a) Using Sachs' analysis, what was the apparent coefficient of friction?

(b) Using the value of μ found in part (a), what value of η would be predicted from Sachs' analysis where $\alpha = 6°$ and $r = 0.4$?

(c) The actual value of η found experimentally with the conditions used in part (b) was 0.80. Explain the discrepancy.

7-3. You are asked to estimate the force required to coin a U.S. quarter, assuming this is done cold. The method approximates axial compression, and the entire quarter flows plastically. The mean flow stress is 25 ksi, the outer diameter is about 0.95 in., and the mean thickness after forming is about 0.060 in. Sticking friction is reasonable.

7-4. The sketch indicates a hot forging of a metal slab from an initial size of 1 in. $\times$ 1 in. $\times$ 10 in. to a final size of $\frac{1}{2}$ in. $\times$ 2 in. $\times$ 10 in. This is accomplished using a flat-faced drop hammer to supply the necessary force, and sticking friction at the interface is a reasonable assumption. For the combination of temperature and strain rate involved, the flow stress of the hot billet is approximately constant at 2500 psi. If yielding follows the von Mises criterion:

(a) Find the *force* required to produce the final thickness.

(b) How much *work* is required to perform this operation?

(c) From what height would a 200 lbm drop hammer have to fall to complete this operation in one blow?

(d) Compute the probable deformation efficiency, η.

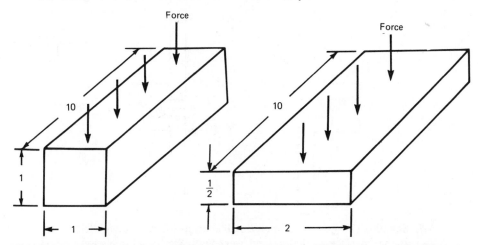

7-5. Two steel plates, whose tensile yield strength, Y, is 100,000 psi, are joined by brazing with a filler metal whose Y is 10,000 psi. Assuming the joint has a uniform thickness, as shown in the sketch, and the bonds between the filler and plates do not break, determine the tensile force, F, needed to cause yielding of this joined unit.

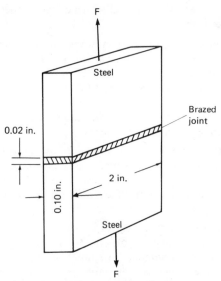

7-6. A thin, lead ring is to be used as a gasket between two steel components as shown in the sketch, which includes all pertinent dimensions. An acceptable seal will result if the gasket is compressed to a final thickness of 0.010 in. The flow stress of the lead is about 2 ksi and work hardening is negligible. Assuming sticking friction at the interface, what force is required to produce the 0.010 in. thickness?

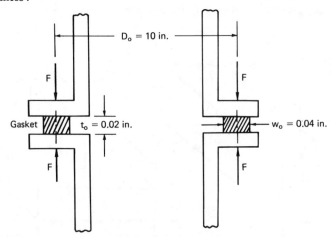

7-7. Equation (7-10) is based upon a slab analysis where Coulomb friction prevailed at the interface of the die and workpiece. Derive an analogous expression if sticking friction, rather than "sliding" friction, prevailed.

7-8. Magnetic permalloy tape is produced by roll flattening of round drawn wire of permalloy. The final as-rolled cross section should be 0.008 × 0.001 in. The rolling direction must be parallel to the wire axis. It is physically possible to achieve the required cross-sectional dimensions with various initial wire diameters, depending on the relative amount of lateral spreading during rolling. It has been found, however, that the best magnetic properties (which depend upon the crystallographic texture formed) result when the rolling is done in such a way as to maximize the amount of spreading.

You are asked for advice on how to design such a rolling process that *maximizes* spreading. Consider the following parameters:

(a) roll diameter, D

(b) reduction per pass, r

(c) friction, μ

(d) back tension, σ_b, and front tension, σ_f

Describe, qualitatively, how you would vary each of these parameters (e.g., using large D, small D, or D doesn't matter) and why. What other parameters might be important?

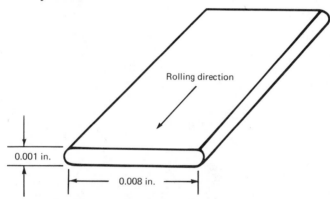

7-9. A tube of 1100-0 aluminum is drawn over a solid, stationary mandrel such that its initial dimensions of 2.435-in. outer diameter and 2.000-in. inner diameter become 2.320 in. and 2.000 in. after flowing through a die having an α of 30°. Since $\bar{\sigma} = 25{,}000\ \bar{\epsilon}^{0.25}$ for this material, a mean yield strength, Y_m, can be found for this operation. If μ is about 0.05 at both the interface of die-tube *and* mandrel-tube, estimate the drawing stress required.

7-10. A metal having a constant flow stress of 5,000 psi is to be drawn from a 4-in. diameter to a 2-in. diameter using a die of 30° semi-angle. Calculate the drawing stress for this situation if the frictional condition at the die-workpiece interface is:

(a) frictionless,

(b) sticking,

(c) $\mu = 0.20$.

7-11. Consider a sheet of metal of 5-in. width and 0.075-in. thickness. It is to be rolled to a thickness of 0.050 in. in one pass using a mill whose steel rolls are of 8-in. diameter; the value of μ in this case is about 0.10, and the average plane-strain flow stress of the metal sheet is 20 ksi.

(a) Calculate the average roll pressure if roll flattening is ignored.

(b) Repeat part (a) if roll flattening is considered.

(c) Estimate the minimum thickness to which this sheet could be rolled.

7-12. Use Eq. (7-14) to predict how the ratio of frictional work to ideal work, w_f/w_i, depends on μ, α, and ϵ_h. (*Hints:* Realize that the derivation of Eq. 7-14 neglects redundant·work, so in effect it assumes $w_r = 0$, or $\sigma_d = w_a = w_i + w_f$. Also, expand the exponential term and after simplifying, assume that ϵ_h is small enough that higher order terms can be neglected.) Describe (in words) how w_f/w_i changes with μ, α, and ϵ_h.

8

Upper-bound analysis

The calculation of the exact loads or forces to cause plastic flow of metals is often difficult, if not impossible. Exact solutions require that both stress equilibrium and a geometrically self-consistent pattern of flow are satisfied simultaneously everywhere throughout the deforming body and on its surface. Fortunately, limit theorems permit force calculations which provide values that are known to be either lower or higher than the actual forces. These calculations provide lower or upper bounds.

A lower-bound solution will give a load prediction that is less than or equal to the exact load needed to cause a body to experience full plastic deformation. As such, it is a "safe" calculation in the design of structures that are not intended to deform plastically. Such an analysis focuses upon satisfying a yield criterion and stress equilibrium, while ignoring the possible shape change of the body. Since this text emphasizes operations where the body *must* deform plastically, no further attention is paid to lower bounds except to note that the ideal work method with $\eta = 1$ provides a lower bound. Several texts [1,2] may be consulted for greater detail on lower bounds.

In metal-forming operations, it is of greater interest to predict a force that will surely cause the body to deform plastically to produce the desired shape change. An upper-bound analysis predicts a load that is at least equal to or greater than the exact load needed to cause plastic flow. Upper-bound analyses focus upon satisfying a yield criterion and assuring that shape changes are geometrically self-consistent. No attention is paid to stress equilibrium.

8-2 UPPER-BOUND ANALYSIS
(kinematically admissible solution)

The upper-bound theorem may be stated as follows: Any estimate of the collapse load of a structure made by equating the internal rate of energy dissipation to the rate at which external forces do work in some assumed pattern of deformation will be greater than or equal to the correct load.

The basis of an upper-bound analysis is:

1. An internal flow field is assumed and must account for the required shape change. As such, the field must be geometrically self-consistent.
2. The energy consumed internally in this deformation field is calculated using the appropriate strength properties of the work material.
3. The external forces (or stresses) are calculated by equating the external work with the internal energy consumption. For a mathematical proof that such solutions predict loads equal to or greater than the exact load to cause plastic deformation, various sources (e.g., [1]) may be consulted.

With such solutions, the assumed field can be checked for complete consistency by drawing a velocity vector diagram, which is commonly called a hodograph.

In applying the upper-bound technique to metal-working operations, several simplifying assumptions are invoked:

1. The work material is isotropic and homogeneous.
2. The effects of strain hardening and strain rate on flow stress are neglected.
3. Either frictionless or constant shear stress conditions prevail at the tool-workpiece interface.
4. Most of the cases considered will be those where the flow is two dimensional (plane strain), with all deformation occurring by shear on a few discrete planes. Elsewhere the material is considered to be rigid. If shear is assumed to occur on intersecting planes that are not orthogonal, these planes cannot, in reality, be planes of maximum shear stress. Many such fields can be posed and the closer such a field is to the true flow field, the closer the upper-bound prediction approaches the exact solution.

8-3 ENERGY DISSIPATION ON A PLANE OF DISCRETE SHEAR

Figure 8-1a shows an element of rigid metal, $ABCD$, moving at unit velocity, V_1, and having unit width into the paper. AD is set parallel to $yy.'$ As the element reaches the plane yy' it is forced to change direction, shape, and velocity. To the right of yy' it now has the shape $A'B'C'D'$ and velocity V_2, at an angle θ_2 to the horizontal. Figure 8-1b is the hodograph; the absolute velocities on

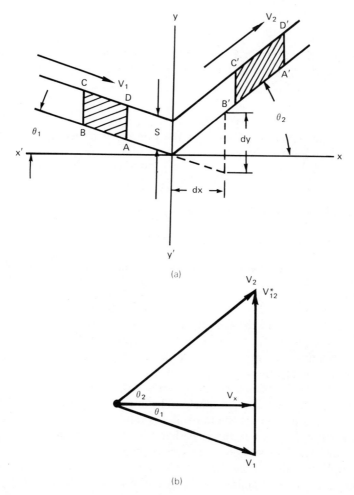

Figure 8-1 (a) Basis for analysis of energy dissipation along a plane of intense shear discontinuity and (b) the hodograph or velocity vector diagram.

either side of yy' are V_1 and V_2 and they are drawn from the origin, O. Both V_1 and V_2 must have the horizontal component, V_x; otherwise, material approaching and leaving yy' would differ in volume; this would violate the concept of incompressibility.

The velocity V_{12}^* is the vector difference between V_1 and V_2 and is the velocity discontinuity along yy'. It is assumed that V_{12}^* occurs along the line (or plane) yy'.

The rate of energy dissipation on yy' must equal the work per volume times the volume per time crossing yy'. Because deformation is due to shear, the work per volume, w, equals the shear stress, τ, times the shear strain, γ. Here, τ must be the shear strength, k, of the metal and $\gamma = dy/dx$, thus,

$$w = k\frac{dy}{dx} \tag{8-1}$$

The volume crossing yy' in an increment of time, dt, is the length of line, S, along yy' times the depth of the plane perpendicular to yy' (unity) times V_x. Thus,

$$\frac{\text{vol}}{\text{time}} = S(1)V_x \tag{8-2}$$

Combining Eqs. (8-1) and (8-2) gives the rate at which work, W, is done to effect this shear deformation;

$$\frac{dW}{dt} = \left(k\frac{dy}{dx}\right)(SV_x) \tag{8-3}$$

Comparing Figs. 8-1a and 8-1b, $dy/dx = V^*_{12}/V_x$ so,

$$\frac{dW}{dt} = kSV^*_{12} \tag{8-4}$$

For deformation fields involving more than one plane of discrete shear,

$$\frac{dW}{dt} = \sum_1^i k \cdot S_i \cdot V^*_i \tag{8-5}†$$

where S_i and V^*_i pertain to each individual plane.

Equation (8-5) is the form we shall use in most problems involving upper bound calculations. It implies that an element deforms in a way that offers maximum plastic resistance.

Most of the flow fields assumed in this text consist of a number of polygons which are viewed as rigid blocks. This means that the velocity of all the material *inside* a polygon is the same and is represented on the hodograph by the point that is common to the lines that bound the polygon on the proposed deformation field. The polygons are separated by the lines of velocity discontinuity, and these discontinuities as well as effects of boundary friction must be considered when summing the contributions to the total internal energy dissipation.

8-4 PLANE-STRAIN FRICTIONLESS EXTRUSION

As an example, consider plane-strain extrusion through frictionless dies as shown in Fig. 8-2. It is only necessary to consider half of the field, where velocity discontinuities occur along AB and BC. Elsewhere, material is undeforming. The hodograph shown in Fig. 8-2b is constructed by drawing a horizontal vector representing the inlet velocity V_o. As the material crosses line AB its velocity is abruptly changed to a new velocity, V_1, which must be drawn parallel

---

†If the plane is curved, $dW/dt = \int k \cdot V^* \cdot dS$ where V^* would vary along dS. This will be of no concern in this text.

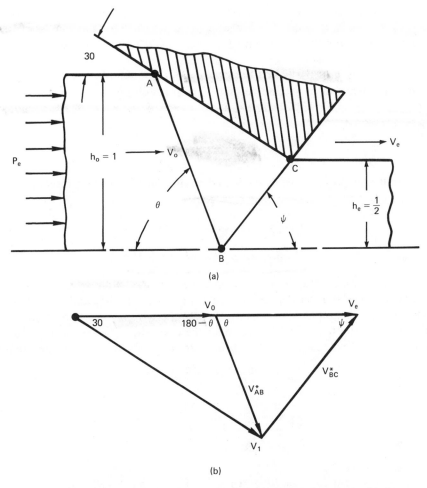

Figure 8-2 (a) Top half of a simple upper-bound field for plane-strain, friction-less extrusion and (b) corresponding hodograph.

to AC since the material in triangle 1 moves as a rigid block. This establishes the direction of V_1. Its magnitude is found by considering the discontinuity velocity V_{AB}^* that a particle suffers when it crosses line AB. This change in velocity corresponds to a shear on plane AB, so V_{AB}^* is represented by a vector parallel to AB and starting at V_o. Since V_1 is the vector sum of the initial velocity, V_o, and the change in velocity V_{AB}^*, this construction establishes the magnitude of V_1.

As a particle in triangle 1 crosses CB, it suffers a second velocity discontinuity, V_{CB}^*, which is represented by a vector parallel to CB. Its magnitude is established by the fact that the resultant final velocity, V_e, must be parallel to

V_o.† A final check on the self-consistency of the hodograph is now possible. The reduction of area requires that the exit velocity V_e equals $V_o(h_o/h_e)$; it does. The hodograph for the lower half of the field could be constructed by exactly the same type of reasoning.

To calculate the extrusion pressure, P_e, let the depth of the slab perpendicular to the plane of projection be unity. The force on this half slab is equal to $P_e(h_o)$ where h_o is the half thickness, and the rate of work required is the force times the velocity, or

$$\frac{dW}{dt} = P_e h_o V_o$$

This input energy is completely dissipated on the planes of velocity discontinuity AB and BC, and this energy balance becomes

$$P_e h_o V_o = k(V^*_{AB}\overline{AB} + V^*_{BC}\overline{BC}) \tag{8-6}$$

or

$$\frac{P_e}{2k} = \frac{1}{2h_o V_o}(V^*_{AB}\overline{AB} + V^*_{BC}\overline{BC}) \tag{8-7}$$

A numerical solution may be obtained by graphically measuring $\overline{AB}$ and $\overline{BC}$ relative to h_o on the physical field, Fig. 8-2a, and V^*_{AB} and V^*_{BC} relative to V_o on the hodograph. Alternatively, an *analytic solution* may be obtained with the use of a little trigonometry. Consider the case where the reduction is 50% and the half die angle, α, is 30°. Then, since $\alpha = 30°$, $\overline{AC} = h_o$; in addition,

$$\sin\theta = \frac{h_o}{\overline{AB}} \quad \text{or} \quad \frac{\overline{AB}}{h_o} = \csc\theta$$

$$\sin\psi = \frac{h_o}{2\overline{BC}} \quad \text{or} \quad \frac{\overline{BC}}{h_o} = \frac{\csc\psi}{2}$$

$$\cot\psi = 2(\cos 30° - \cot\theta)$$

$$\frac{V^*_{AB}}{\sin 30} = \frac{V_o}{\sin(\theta - 30)} \quad \text{or} \quad \frac{V^*_{AB}}{V_o} = \frac{\sin 30}{\sin(\theta - 30)}$$

$$\frac{V^*_{BC}}{\sin\theta} = \frac{V^*_{AB}}{\sin\psi}$$

so

$$\frac{V^*_{BC}}{V_o} = (\sin\theta \csc\psi)\frac{V^*_{AB}}{V_o}$$

It is obvious that the magnitude of $P_e/2k$, for this field, will depend upon the value selected for θ. Consider the case where $\theta = 90°$, thus $\psi = 30°$. Here,

$$V^*_{AB} = V_o\left[\frac{\sin 30}{\sin(90 - 30)}\right] = 0.577V_o$$

---

†If the die were asymmetric, V_e would not necessarily be parallel to V_o. It would be necessary, however, that the final velocity, V_e, predicted by the upper half of the field be the same as that predicted by the lower half. In the case of a symmetric die and deformation field, this condition corresponds to V_e being parallel to V_o.

$$V_{BC}^* = (\sin 90)(\csc 30)(0.577V_o) = 1.154V_o$$

$$\overline{AB} = h_o, \quad \overline{BC} = \csc 30 \frac{h_o}{2} = h_o$$

Thus, $\dfrac{P_e}{2k} = \dfrac{1}{2h_oV_o}[0.577V_oh_o + 1.154V_oh_o] = 0.8655$ (8-8)

As θ, and thus ψ, are varied, the value of $P_e/2k$ would vary as shown in Fig. 8-3; the lowest value of ≈ 0.78 occurs when $\theta \approx 72°$. Note that a lower bound can readily be found from the homogeneous work method discussed earlier, where

$$P_e = \int \sigma d\epsilon = 2k \ln(2) = 1.386k$$

or $\dfrac{P_e}{2k} = 0.693$ (8-9)

As will be shown in Chap. 9, an exact solution for this problem gives $P_e/2k = 0.762$ for frictionless conditions.

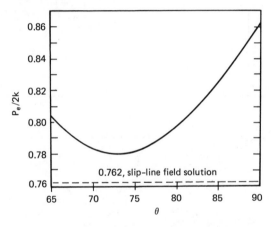

Figure 8-3   Variation of extrusion pressure with θ for the upper-bound field in Fig. 8-2.

Again, it is not essential to assume frictionless conditions. A simple alternative assumption is that sticking friction prevails along the die face. If, for the flow field shown in Fig. 8-2a, sticking friction occurs along AC, there must be a shear stress, k, along AC, and V_1 must be considered as a velocity discontinuity. An additional term, $k(V_1/V_o)(\overline{AC}/h_o)$, must be added to the equation for the upper bound. Then,

$$\frac{P_e}{2k} = \frac{\frac{1}{2}(V_1 \cdot \overline{AC} + V_{BC}^* \cdot \overline{BC} + V_{AB}^* \cdot \overline{AB})}{V_oh_o}$$

$$= \frac{P_e}{2k}(\text{frictionless}) + \frac{1}{2}\left(\frac{V_1}{V_o}\right)\left(\frac{\overline{AC}}{h_o}\right)$$ (8-10)

A plot of calculated values of $P_e/2k$ versus θ for this condition gives the minimum solution ($P_e/2k = 1.432$) at $\theta \approx 83°$.

More complex deformation fields may be postulated to obtain still lower values of an upper bound. Consider the half-field of Fig. 8-4a for the same problem; the corresponding hodograph is shown in Fig. 8-4b. These give an upper bound of $P_e/2k = 0.768$. Still lower values can be obtained by modifying this two-triangle half-field, in the same manner as discussed with the one-triangle field.

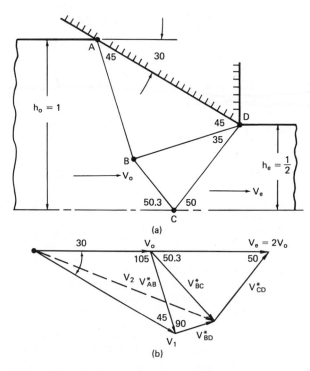

Figure 8-4   (a) Top half of a two-triangle upper-bound field for plane-strain frictionless extrusion and (b) corresponding hodograph.

The internal distortion corresponding to the proposed field can be determined by constructing a vertical reference, PR, in Fig. 8-5a, at the die inlet and finding the positions of particles after an elapsed time, t. Here, t is long enough for all such particles to have emerged from the field. First, the flow paths of several particles are constructed, noting that directions of the paths in each region of the field must be parallel to the corresponding velocity vector on the hodograph, Fig. 8-5b, The total time, t, is the sum of the times spent in each region,

$$t = t_o + t_1 + \ldots + t_e \quad \text{or} \quad t_e = t - \sum t_i \qquad (8\text{-}11)$$

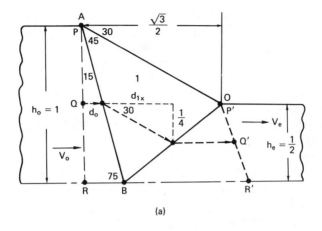

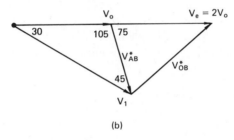

(b)

Figure 8-5 (a) A single triangle field and (b) hodograph used for analyzing grid distortion. Here, uniform shear results.

where t_o and t_e are the times spent before entering and after leaving the field. Other subscripts refer to regions in the field. Substituting $t = d/V$,

$$t_e = \frac{d_e}{v_e} = t - \sum (d_i/V_i) \tag{8-12}$$

so $$d_e = V_e[t - \sum (d_i/V_i)] \tag{8-13}$$

The total distance traveled, $d = d_e + \sum d_i$, so

$$d = V_e t + \sum d_i \left(1 - \frac{V_e}{V_i}\right) \tag{8-14}$$

Calculations are often simplified if the horizontal components of the velocities and directions are used in Eq. (8-14) to give

$$d_x = V_e t + \sum d_{ix} \left(1 - \frac{V_e}{V_{ix}}\right) \tag{8-15}$$

To illustrate this procedure, consider the field in Fig. 8-5 for a 50% reduction with $\alpha = 30°$. Let $h_o = 1$ and $V_o = 1$, so $V_e = 2$. Point P on the surface travels through the field along AO so $d_o = 0$, $d_{1x} = \sqrt{3}/2$, $V_{1x} = $ (sin 105°/sin 45°) cos 30° = 1.183. Now, $d_x = 2t + (\sqrt{3}/2)(1 - 2/1.183) = 2t - 0.598$. If t is chosen as one, $d_x = 1.402$. For point Q, mid-way between surface and centerline, $d_o = \tan 15°/2 = 0.134$, $d_{1x} = \sqrt{3}/4$, and $V_{1x} = 1.183$ as above, so $d_x = (2)(1) + 0.134(1 - 2) + \sqrt{3}/4(1 - 2/1.183) = 1.567$.

For point R on the centerline, $d_o = \tan 15° = 0.268$, $d_1 = 0$, so $d_x = 2 - 0.268 = 1.732$. New positions of these points, P', Q', and R' fix the new position of the grid line in Fig. 8-5a and indicate the distortion caused by this field. This implies that the material has undergone a uniform net shear parallel to V_o, this shear reflecting the redundant strain. A key observation is that a line such as $P'Q'R'$ is straight because every particle on PQR experiences the same strain history.

Next consider the situation shown in Fig. 8-6 using the same reduction but a two-triangle field. Here, a particle crossing BC is first subjected to a velocity jump V_{BC}^* and, subsequently, V_{OC}^*, so as to exit with velocity V_e. A particle crossing AB suffers three shear discontinuities, V_{AB}^*, V_{OB}^*, and V_{OC}^* before exiting with velocity V_e. As a consequence, the strain history and, in essence, the time, of particles entering the field above and below B is not the same. This leads to a strain gradient as shown by line $P'Q'R'$ in Fig. 8-6. If the best upper bound is found for the one- and two-triangle fields, the latter predicts the single lowest value of $P/2k$. This implies that the existence of strain gradients requires less

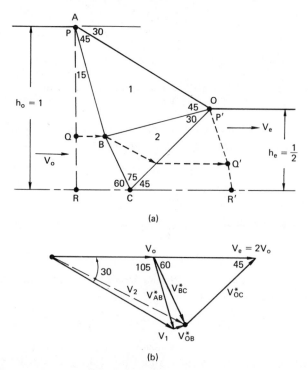

(a)

(b)

Figure 8-6 (a) A two-triangle field and (b) hodograph used for analyzing grid distortion. Here, non-uniform shear results.

work to effect the same shape change. Such gradients are observed in many deformation processes.

8-5 PLANE-STRAIN INDENTATION
WITH FRICTIONLESS INTERFACE

As a second example, in Fig. 8-7 the deformation field is assumed to be composed of equilateral triangles. Due to symmetry, only half the field need be considered.

As the punch moves downward, shear occurs along the lines OA, AB,

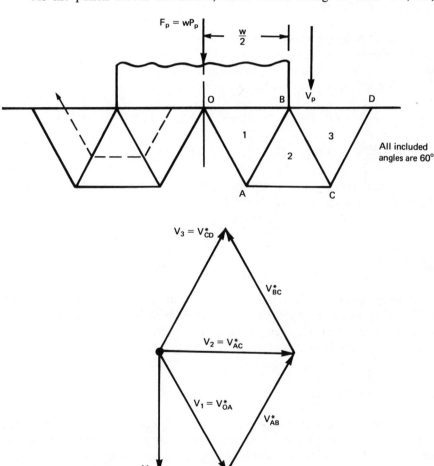

Figure 8-7 An upper-bound field for plane-strain, frictionless indentation of a semi-infinite body and the hodograph for the right half of the field.

AC, BC, and CD. Frictionless conditions prevail at the interface, so no shear occurs along line OB. The metal below or outside of the triangles 1, 2, 3, is assumed to be rigid and stationary, while the punch velocity is V_p. The absolute velocity of metal in triangle 1 is V_1 and it must be parallel to OA along which shearing occurs. V_1 is parallel to OA and has a vertical component equal to V_p. The velocity discontinuity, V^*_{OA}, is equal to V_1. Note that because of the frictionless condition on OB, the hodograph does not show V^*_{OB}. The velocity in region 2 must be parallel to line AC along which shear occurs, therefore $V_2 = V^*_{AC}$. There is also a velocity discontinuity, V^*_{AB} along line AB, which is the velocity difference between V_1 and V_2. Similarly, the exit velocity $V_3 = V^*_{CD}$ is parallel to line CD and occurs because of a velocity discontinuity across BC, V^*_{BC}.

The hodograph shows that the five velocity discontinuities are equal; their magnitude is $V_p/\cos 30° = 2V_p/\sqrt{3}$. In addition, the lengths of the five lines of shear are the same; they equal $w/2$.

With Eq. (8-5), the internal rate of energy consumption is

$$\frac{dW}{dt} = k(\overline{OA}V^*_{OA} + \overline{AB}V^*_{AB} + \overline{AC}V^*_{AC} + \overline{BC}V^*_{BC} + \overline{CD}V^*_{CD})$$

or $$\frac{dW}{dt} = 5k\left(\frac{w}{2} \cdot \frac{2V_p}{\sqrt{3}}\right) = \frac{5}{\sqrt{3}}kwV_p$$

The external energy applied by the punch for the right half of the field is $P_p(w/2)V_p$. Equating this with $5kwV_p/\sqrt{3}$ gives

$$P_p w \frac{V_p}{2} = \frac{5kwV_p}{\sqrt{3}}$$

or $$\frac{P_p}{2k} = \frac{5}{\sqrt{3}} = 2.89 \tag{8-16}$$

Other upper bounds could be used and, in the limit, the best one would approach $P/2k = 2.57$ which, as we shall see, is an exact solution. If sticking friction occurred along line OB, then an extra term would be added to the energy dissipation; it would be $k \cdot w/2 \cdot V_1 = k(w/2)(V_p/\sqrt{3})$.

8-6 ANOTHER APPROACH TO UPPER BOUNDS

To this point, the deformation fields for upper-bound analyses have consisted of rigid regions separated by planes upon which discrete shear occurs. However, other deformation fields are suitable as long as they are kinematically admissible. For example, an acceptable field may include regions undergoing homogeneous deformation. In the sections that follow, it will be shown that the fields assumed for plane-strain and axisymmetric compression, with either sticking friction or a constant shear stress interface, are suitable for upper-bound analyses. In addition, the solutions for these two situations, as determined by a force-balance slab analysis as covered in Chap. 7, are also upper bounds. For the cases of plane-strain or axisymmetric drawing, velocity discontinuities at the die entrance and exit introduce an additional term in a correct upper-bound solution. The following sections are based on reference [3].

8-7 PLANE-STRAIN COMPRESSION WITH STICKING FRICTION

Consider an element of dimensions $h\,dx$ at a distance, x, from the centerline as shown in Fig. 8-8. The element is assumed to deform homogeneously as it slides away from the centerline. Instead of using a force balance, as discussed in Chap. 7, we will calculate the rate of internal energy dissipation and equate that to the external work rate. Homogeneous deformation requires an energy dissipation of $h(dx)\sigma_o\dot{\epsilon}$. With $\sigma_o = 2k$ and $\dot{\epsilon} = 2V_o/h$, this work rate is

$$\dot{W}_h = 4kV_o\,dx \tag{8-17}$$

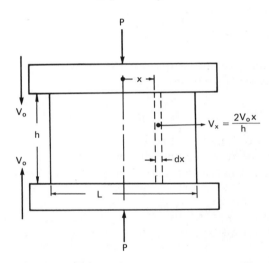

Figure 8-8 Essentials for making a slab-energy analysis of plane-strain compression with sticking friction.

An admissible velocity field, which satisfies volume constancy, gives the horizontal velocity component of this element as

$$V_x = 2V_o\frac{x}{h} \tag{8-18}$$

The rate of energy dissipation due to sliding at both interfaces is, using Eq. (8-18),

$$2kV_x\,dx = 4kV_o\left(\frac{x}{h}\right)dx \tag{8-19}$$

Equating the external work rate, $2PLV_o$, with the *total* rate of internal energy dissipation as given by Eqs. (8-17) and (8-19) gives

$$2PLV_o = 2 \int_o^{L/2} 4kV_o \left(1 + \frac{x}{h}\right) dx$$

or

$$\frac{P}{2k} = 1 + \frac{L}{4h} \qquad (8\text{-}20)$$

where P is the average pressure.

This solution is *identical* to that obtained using a standard force-balance slab analysis as shown by Eq. (7-25). This derivation satisfies all upper-bound requirements, and is therefore a valid upper bound. Note that if a constant interface shear stress, mk, had been used, the solution would have been

$$\frac{P}{2k} = 1 + \frac{mL}{4h} \qquad (8\text{-}21)$$

which is mentioned in Sec. 7-8.

8-8 DISC COMPRESSION

A similar slab upper-bound analysis for axisymmetric disc compression with constant interface shear stress, mk, gives

$$\frac{P}{2k} = \frac{\sigma_o}{2k} + \frac{mR}{3h} \qquad (8\text{-}22)$$

This is identical with Eq. (7-35) if m is taken as unity, that is, for sticking friction. Because there are no internal discontinuities, in the two problems just discussed, the force-balance and energy approaches provide identical answers!

8-9 TRADITIONAL UPPER BOUNDS
FOR PLANE-STRAIN COMPRESSION

For plane-strain operations, as shown by earlier examples, the usual approach to upper bounds is to imagine that all deformation occurs along discrete lines (actually planes in three dimensions). One such field for this problem is shown in Fig. 8-9a, where the regions AOD and COB are dead-metal zones that move with the same velocity, V_o, as the compression platens. Discrete shear occurs on lines AO, BO, CO, and DO and the corresponding hodograph is shown in Fig. 8-9b. Equating the rate of external work to internal energy dissipation gives

$$2PLV_o = 4k\overline{AO}V_{AO}^* \qquad (8\text{-}23a)$$

Substituting $\overline{AO} = \frac{1}{2}(h^2 + L^2)^{1/2}$ and $V_{AO}^* = V_o(h^2 + L^2)^{1/2}/h$, the upper bound is

$$\frac{P}{2k} = \frac{1}{2}\left(\frac{h}{L} + \frac{L}{h}\right) \qquad (8\text{-}23b)$$

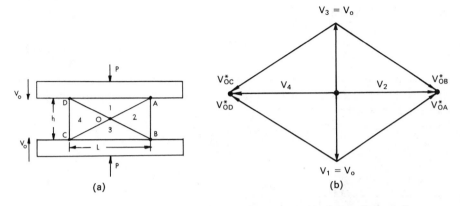

(a)

(b)

Figure 8-9 (a) A proposed upper-bound field and (b) the full hodograph for plane-strain compression with sticking friction. Reprinted by permission of the Council of the Institute of Mechanical Engineers. From R. M. Caddell and W. F. Hosford, *Int. J. Mech. Eng. Educ.*, 8 (1980).

For large values of L/h, better upper-bound solutions are obtained with fields consisting of more than one triangle along the work metal-platen interface, the lowest solutions corresponding to an odd number (3, 5, . . .) of triangles. Only with an odd number is there a dead-metal cap in the center which does not slide against the platens. The field consisting of three triangles and the corresponding hodograph for the upper right quarter of the field are shown in Fig. 8-10. The general solution is

$$\frac{P}{2k} = \frac{3h}{2L} + \frac{L}{2h} + \frac{w^2}{2hL} - \frac{w}{2h} \tag{8-24}$$

where w is the base of the center triangle. The lowest value of $P/2k$ occurs

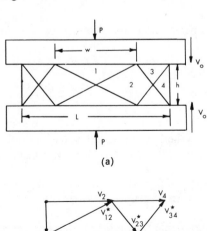

(a)

Figure 8-10 Same as Fig. 8-9 using a different upper-bound field. From R. M. Caddell and W. F. Hosford, *ibid.*

(b)

when $w = L/2$, and is

$$\frac{P}{2k} = \frac{3h}{2L} + \frac{3L}{8h} \qquad (8\text{-}25)$$

A field of five triangles and the corresponding hodograph are shown in Fig. 8-11. Here it is assumed that $\overline{AB} = \overline{BC}$. With this assumption, the lowest value occurs when $w = L/3$, and is

$$\frac{P}{2k} = \frac{5h}{2L} + \frac{L}{3h} \qquad (8\text{-}26)$$

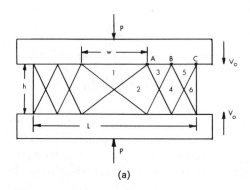

(a)

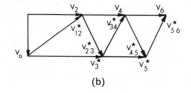

(b)

Figure 8-11 Same as Figs. 8-9 and 8-10 using still another field. From R. M. Caddell and W. F. Hosford, *ibid.*

A general minimum solution for this class of upper bounds is

$$\frac{P}{2k} = \frac{1}{hL}\left[\frac{nh^2}{2} + C\left(\frac{L}{n+1}\right)^2\right]$$

where

$$C = \frac{3n+1}{2} + \sum_{i=1}^{i=(n-1)/2}(2i-1) \qquad (8\text{-}27)$$

and n is an odd integer ≥ 3. This minimum occurs when $w = 2L/(n+1)$.

Whether still lower values would occur with this class of field when the ratio of $\overline{AB}/\overline{BC}$ is varied has not been investigated.

8-10 A COMBINED APPROACH

It is possible to combine the upper-bound slab and the traditional upper-bound analyses to take advantage of the dead-metal cap. Such a field is shown in Fig. 8-12, together with the hodograph for the upper right portion of the central

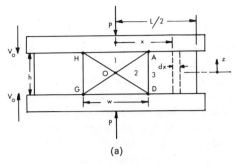

(a)

Figure 8-12 (a) Combined conventional upper-bound and slab-energy fields and (b) a partial hodograph for plane-strain compression with sticking friction. From R. M. Caddell and W. F. Hosford, *ibid*.

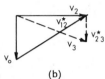

(b)

region. The central region is a simple upper-bound field of width, w, while the remainder of the workpiece is assumed to undergo homogeneous deformation. The rate of energy dissipation along AO, OD, OG, and OH is

$$\frac{2kV_o(h^2 + w^2)}{h} \qquad (8\text{-}28)$$

Energy is also dissipated along AD and GH. The velocity discontinuity, which varies along these lines, increases with vertical distance, z, from the centerline, as $V_{23}^* = 2zV_o/h$. The total energy dissipated along AD and GH is given by

$$4 \int_0^{h/2} kV_{23}^* \, dz = khV_o \qquad (8\text{-}29)$$

The analysis for the outer region is similar to that for the slab upper bound. For constant volume, the horizontal velocity for $x > w/2$ is $V_x = 2V_ox/h$, so the total energy dissipated along the platens is

$$4 \int_{w/2}^{L/2} kV_x \, dx = \frac{kV_o(L^2 - w^2)}{h} \qquad (8\text{-}30)$$

The homogeneous deformation between $w/2 \le x \le L/2$ occurs with a strain rate $\dot{\epsilon} = 2V_o/h$, so the energy dissipation rate is

$$2k\dot{\epsilon}(L - w)h = 4kV_o(L - w) \qquad (8\text{-}31)$$

Equating the external energy rate, $2PLV_o$, with the dissipation given by the sum of Eqs. (8-28), (8-29), (8-30), and (8-31) gives

$$\frac{P}{2k} = 1 + \frac{3h}{4L} + \frac{L}{4h} - \frac{w}{L} + \frac{w^2}{4hL}$$

The minimum occurs when $w = 2h$, and for this geometry,

$$\frac{P}{2k} = 1 + \frac{L}{4h} - \frac{h}{4L} \qquad\qquad (8\text{-}32)$$

Of course this solution is valid only when $L > 2h$.

It is interesting to note that $w = 2h$ corresponds to a size of dead-metal cap and to local velocities that are quite similar to those predicted by a slip-line field shown by Johnson and Mellor [1]. This is probably why the combined field gives a low upper bound.

Values of $P/2k$ predicted for a range of L/h values from the various analyses above and from a slip-line field analysis are plotted in Fig. 8-13. It can be seen that the new combined slab-traditional upper bound predicts a *continuous* function which gives a low solution for a wide range of L/h.

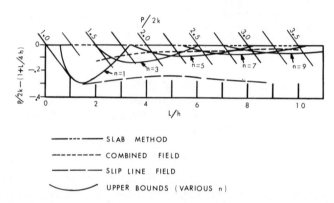

Figure 8-13 Plot of $P/2k$ versus L/h using results from analyses based upon fields in Figs. 8-9 through 8-12 plus results from a slip-line field from reference [1]. The unusual coordinate system allows a comparison of the various solutions. On cartesian coordinates, the solutions are too close to one another to distinguish easily. From R. M. Caddell and W. F. Hosford, *ibid.*

8-11 PLANE-STRAIN WEDGE OR SHEET DRAWING

Consider the plane-strain drawing of a sheet from an initial thickness, t_o, to a final thickness, t_f, using a wedge-shaped die of semi-angle α; a constant shear stress, mk, prevails at the interface. Figure 8-14a shows the flow lines and Fig. 8-14b, the hodograph. The paths of all particles are horizontal before crossing AA' and after exiting across BB'. Between these inlet and outlet boundaries, the flow lines are straight and converging. All particles on a common vertical line such as CC' have a common horizontal velocity component, V_x, but such velocities constantly increase from AA' to BB'. Upon crossing AA', a particle suffers a velocity discontinuity, V_A^*, that is proportional to its distance, y, from the centerline. At the surface, where $y = t_o/2$, the velocity of a particle abruptly changes from V_o to $V_{S(A)}$ due to the velocity discontinuity $V_{S(A)}^* = V_o \tan \alpha$. For a particle on flow line Q, at a distance, y, from the centerline, the velocity

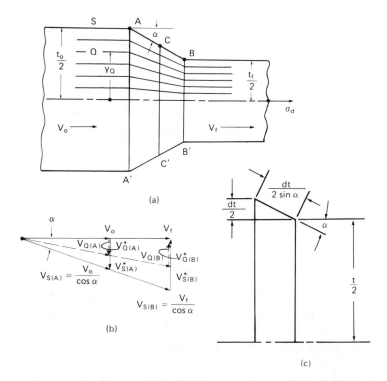

Figure 8-14 (a) Flow lines, (b) a partial hodograph, and (c) an element in the deformation zone used for predicting the extrusion pressure or drawing stress for plane-strain deformation assuming constant shear stress, mk, at the interface.

discontinuity along AA' is $V^*_{Q(A)} = yV_o \tan \alpha/(t_o/2)$. The rate of energy dissipation over a differential height, dy, at any arbitrary y is $kdy(yV_o \tan \alpha)/(t_o/2)$, so over the entire section along AA',

$$\dot{W}_r = 2 \int_0^{t_o/2} \frac{kV_o \tan \alpha}{t_o/2} y\,dy = \frac{kV_o t_o (\tan \alpha)}{2} \qquad (8\text{-}33)$$

At the exit, BB', an identical dissipation term is found, so the total rate of energy dissipated along these two discontinuities is

$$\dot{W}_r = kV_o t_o \tan \alpha \qquad (8\text{-}34)$$

Next consider the frictional energy expended along the die interface by taking a slab whose thickness is t at some arbitrary position in the deformation zone, as shown in Fig. 8-14c. The contact area, including both faces, is $dt/\sin \alpha$. Constancy of volume requires that the horizontal velocity component associated with this slab be $V_x = V_o t_o/t$, so the velocity of metal moving along the surface is

$$V_S = V_x/\cos \alpha = \frac{V_o t_o}{t \cos \alpha} \qquad (8\text{-}35)$$

The rate of energy dissipation at the interface is then related to

$(mk)(V_o t_o/t \cos \alpha)(dt/\sin \alpha)$ such that

$$\dot{W}_f = \int_{t_f}^{t_o} \frac{m}{\cos \alpha} \frac{kV_o t_o}{\sin \alpha} \frac{dt}{t} = \frac{mkV_o t_o}{\cos \alpha \sin \alpha} \epsilon \tag{8-36}$$

where

$$\epsilon = \ln \left(\frac{t_o}{t_f} \right)$$

The rate of energy dissipation due to homogeneous deformation is

$$\dot{W}_h = t_o V_o \sigma_o \epsilon = 2kt_o V_o \epsilon \tag{8-37}$$

Now the external work rate, $\sigma_d V_f t_f = \sigma_d V_o t_o$, is equated with the results in Eqs. (8-34), (8-36), and (8-37) to give

$$\dot{W}_a = \dot{W}_h + \dot{W}_f + \dot{W}_r \tag{8-38}$$

The result is

$$\sigma_d/2k = \left(1 + \frac{m}{\sin 2\alpha} \right) \epsilon + \frac{1}{2} \tan \alpha \tag{8-39}$$

This can be interpreted as being related to the work balance, $\sigma_d = w_h + w_f + w_r$, where the homogeneous work, $w_h/2k = \epsilon$, the frictional work, $w_f/2k = m\epsilon/\sin 2\alpha$, and the redundant work, $w_r/2k = \frac{1}{2} \tan \alpha$.

It is noted that the use of a force balance slab method produces Eq. (8-39) without the redundant work term of $\frac{1}{2} \tan \alpha$. This can be shown by using the term mk in place of the friction term μP in Eq. (7-1) and following the derivation that led to Eq. (7-10). Because the redundant work term is neglected in such a force balance, that result is not a true upper bound.

8-12 AXISYMMETRIC ROD DRAWING

Consider the drawing of a rod from an initial diameter of D_o to a final diameter of D_f through a die of semi-angle α with a constant shear stress, mk, at the interface. The procedure parallels that in Sec. 8-11 except for the difference in geometry.

A slab upper bound can be obtained by equating the external work rate to the rate of internal energy dissipation. In a slab element of radius R, the horizontal component of velocity is $V_x = V_o R_o^2/R^2$, so the velocity of sliding at the die interface is $V_x/\cos \alpha = V_o R_o^2/(R^2 \cos \alpha)$. The area of the element in contact with the die is $2\pi R dR/\sin \alpha$ and the shear stress at the interface is mk. Thus the total rate of frictional work over the whole die surface is

$$\dot{W}_f = \int_{R_f}^{R_o} \frac{mkV_o R_o^2 2\pi R dR}{R^2 \cos \alpha \sin \alpha} = \frac{2\pi mk R_o^2 V_o}{\sin 2\alpha} \epsilon \tag{8-40}$$

Energy is also dissipated because of the velocity discontinuities at the inlet and outlet of the die. On entering the die a particle suddenly develops a radial component of velocity V_R^* which depends upon its distance, y, from the centerline. Since $V_R^* = V_o(y/R_o) \tan \alpha$, the rate of energy dissipation at the inlet is

$$\dot{W}_r = \int_0^{R_o} kV_o\left(\frac{y}{R_o}\right)(\tan \alpha)2\pi y \, dy = \frac{2}{3}kV_o\pi R_o^2 \tan \alpha \qquad (8\text{-}41)$$

and the term at the die exit is exactly the same.

The homogeneous work rate is

$$\dot{W}_h = \pi R_o^2 \sigma_o V_o \epsilon = \pi R_f^2 \sigma_o V_f \epsilon \qquad (8\text{-}42)$$

Equating the external work rate, $\sigma_d \pi R_f^2 V_f = \sigma_d \pi R_o^2 V_o$, to the sum of Eqs. (8-40), (8-42), and twice Eq. (8-41), and solving for $\sigma_d/2k$,

$$\frac{\sigma_d}{2k} = \left(\frac{\sigma_o}{2k} + \frac{m}{\sin 2\alpha}\right)\epsilon + \frac{2}{3}\tan \alpha \qquad (8\text{-}43)$$

Again, a force-balance slab analysis produces Eq. (8-43) without the redundant work term, $\frac{2}{3}\tan \alpha$, and in that case, the result is not a true upper bound.

Other kinematically admissible fields may be used to analyze the type of problems just discussed. As an example, Avitzur [4] has derived an upper bound for axisymmetric drawing that predicts drawing stresses slightly lower than those from Eq. (8-43). His velocity field leads to more complex mathematical developments, and although the result is a better upper bound, the difference between predictions based upon Eq. (8-43) and his work is quite minimal.

8-13 GENERAL OBSERVATIONS

From these previous examples it has been shown that, for plane-strain and axisymmetric compression, the slab method based upon a force balance does satisfy all requirements of the upper-bound theorem if sticking friction or a constant interfacial shear stress prevails. In most situations this is not true, since redundant deformation cannot be neglected. It seems useful to separate the different contributions to upper-bound predictions into, (homogeneous) + (boundary friction) + (redundant deformation) in order to portray the physical impact of each.

REFERENCES

[1] W. Johnson and P. B. Mellor, *Engineering Plasticity*. New York: Van Nostrand Reinhold, 1973, pp. 392–401, 416–17, 418–19.

[2] C. R. Calladine, *Engineering Plasticity*. Elmsford, N.Y.: Pergamon Press, 1969, pp. 94–104.

[3] R. M. Caddell and W. F. Hosford, *Int. J. Mech. Eng. Ed.* 8 (1980), pp. 1–6.

[4] B. Avitzur, *Metal Forming: Processes and Analysis.* New York: McGraw-Hill, 1968, pp. 153–64.

PROBLEMS

8-1. For the conditions related to Fig. 8-2, determine $P_e/2k$ if θ is 80° and compare your answer with the plot on Fig. 8-3.

8-2. Plane-strain frictionless drawing is to be analyzed for the conditions shown in the sketch; only the top half of the physical field is shown. All included angles in triangles ABC and CDE are 60°, while AB and CD are perpendicular to the centerline. Calculate $\sigma_d/2k$ for this situation, and comment on the physical implication of your answer.

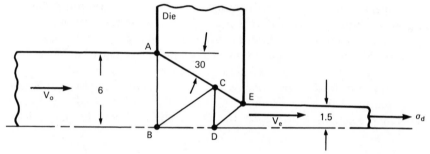

8-3. A flat, frictionless punch is to indent a metal under plane-strain conditions. For the upper-bound field shown, draw a hodograph to scale and find $P/2k$.

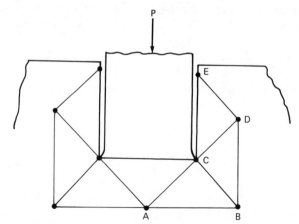

8-4. Plane-strain compression, using a proposed upper-bound field, is shown in the sketch; sticking friction occurs along *AD* and *BC*. Calculate *P/2k* for *L/H* values of 1, 2, 3, 4, and 5 respectively. Compare your findings with Fig. 8-13.

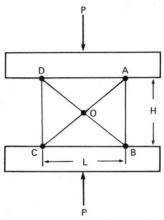

8-5. Indenting, as shown in Fig. 8-7, where all included angles were 60°, predicted $P/2k = 2.89$ as shown by Eq. (8-16). If the angles *OAB*, *ABC*, and *BCD* were 90° and the other angles were 45°, find *P/2k* and compare it with the value of 2.89.

8-6. A proposed upper bound for extrusion is shown in the sketch. For this half-field, draw a hodograph to scale and determine the absolute velocity of particles inside the region bounded by lines 6, 7, and 8. Use the scaled drawing as accurately as possible.

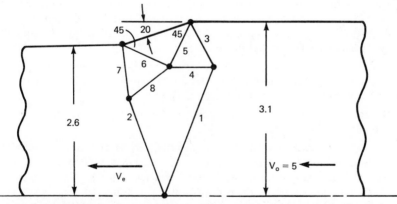

8-7. The sketch illustrates plane-strain extrusion involving two dead-metal zones inside *ADB* and *FEG*. For the upper-bound field shown,
   (a) Calculate $P_e/2k$.
   (b) Determine the velocity inside *ABC*.
   (c) Determine V_{AC}^*.
   (d) Compute a deformation efficiency for this condition.

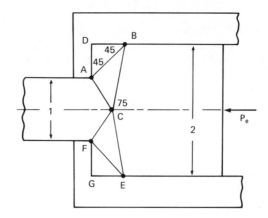

8-8. Repeat Prob. 8-7 for an upper-bound field where BC and CE are perpendicular to the centerline and AC and FC are adjusted accordingly to maintain single triangles ABC and CFE.

8-9. Reanalyze Prob. 8-4 if frictionless conditions prevailed along AD and BC and compare these answers with those in Prob. 8-4.

8-10. For the situation in Prob. 8-4, use the slab-analysis solution given by Eq. (7-25) to analyze the situation for the five values of L/H stated.
(a) Plot the results of these two problems to a common scale of $P/2k$ versus L/H.
(b) Is the slab method an upper-bound solution? Explain.

8-11. Refer to Fig. 8-5 and consider two points U and V on line PQR that are mid-way between P and Q, and Q and R, respectively. Following the procedures connected with Eq. (8-15) that indicated the relative position of $P'Q'R'$, calculate the appropriate distances traveled in time $t = 1$ to locate U' and V' on line $P'Q'R'$. Then, to some convenient scale, plot points P', U', Q', V', and R' to convince yourself a straight line results.

8-12. For the field given in Fig. 8-10 and assuming sticking friction at the platen-workpiece interface, complete a full derivation to produce Eq. (8-24). Then show that Eq. (8-25) is the minimum value of $P/2k$.

8-13. Repeat Prob. 8-12 for the field shown in Fig. 8-11 to prove that Eq. (8-26) provides a minimum value for $P/2k$.

8-14. For axisymmetric rod drawing, with a constant interfacial shear stress of $0.1k$, a reduction, r, of 30%, semi-die angle of 10°, and a material flow stress of 20,000 psi, calculate the drawing stress required according to Eq. (8-43). Assume the Tresca criterion applies. If the von Mises criterion were used, what would be the predicted value of σ_d? Determine the deformation efficiency for both cases.

8-15. Two different upper-bound fields for a 2:1 reduction by extrusion are shown in the sketch. Regions ABC and EFG are dead-metal zones, and the line JG is parallel to the centerline.
(a) Calculate $P/2k$ for each field.
(b) Determine the deformation efficiency, η, for each field.
(c) What is the absolute velocity of a particle in region JGH?

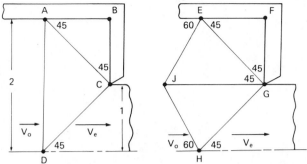

8-16. Plane-strain, frictionless extrusion for a 6 : 1 reduction is to be analyzed using the upper-bound field shown; only the top half is indicated. Draw a hodograph to scale, find $P_e/2k$, and determine the velocity of a particle in *BDC*.

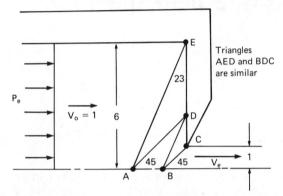

Triangles AED and BDC are similar

8-17. Asymmetric, frictionless, plane-strain extrusion is to be analyzed using the upper-bound field shown.
   (a) Draw the complete hodograph to scale to show that the same exit velocity must be common to both the top and bottom regions of the field.
   (b) Determine the angle, θ, between the extruded material and the *x*-axis.
   (c) On which shear discontinuity is the largest amount of energy expended?

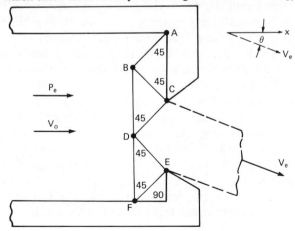

9

Slip-line field theory

9-1 INTRODUCTION

In the discussion of the upper-bound approach as it is applied to plane-strain problems, deformation was assumed to occur along lines of intense shear. Metal inside any polygon moved as a solid block as deformation took place. In an attempt to include a more realistic picture of actual metal flow in the analyses of metal deformation, a type of field theory has been developed. This can be viewed as a graphical approach, which portrays the flow pattern from point to point in the deforming metal, and is referred to as slip-line field analysis. To some extent this nomenclature is unfortunate since, to the metallurgist, the term "slip line" has a meaning of its own. Perhaps the name "shear-plane field theory" might have been better, since slip lines refer to planes of maximum shear which are oriented at 45° to principal planes. There should be no confusion on this point if it is remembered that the "slip lines" denote the planes of maximum shear stress.

This analysis is based upon a deformation field that is geometrically consistent with the shape change. Furthermore, the stresses within the field are statically admissible. However, because the stress state outside the field is ignored, some proposed fields, which are otherwise acceptable, may violate equilibrium outside the deformation zone; in this event, such solutions are upper bounds.

Besides the usual assumptions that the metal is isotropic and homogeneous, the *common* approach to this subject usually involves the following:

1. The metal is rigid-perfectly plastic; this implies the neglect of elastic strains and treats the flow stress as a constant (no work hardening).
2. Deformation is by plane strain.
3. Possible effects of temperature, strain rate, and time are not considered.
4. There is a constant shear stress at the interfacial boundary. *Usually*, either a frictionless condition or sticking friction is assumed.

Prior to the development of the essential groundwork that underlies this theory, a few points are worth considering.

1. The approach taken in this text is not intended to be all inclusive in regard to its mathematical development. Other sources [1, 2] can be consulted for a more rigorous attack. Instead, only the essential features are presented.

2. Except in a few cases involving simple fields, the reader is not expected physically to construct a slip-line field.† What is intended is that the basic theory and a sufficient number of examples will be presented so that an understanding of this technique will be grasped. If one intends to apply this theory to new problems, a proper literature search [4], practice, and ingenuity are essential.

As important as it is to understand the mathematical framework that will follow, one should not lose sight of the physical concepts covered by this technique. Towards this end, consider the operation shown in Fig. 9-1.

Two flat anvils are loaded to compress a sheet or plate of width, *w*, and

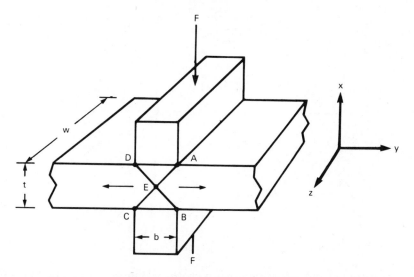

Figure 9-1   A slip-line field for frictionless indentation, where $t = b$.

†For those interested, reference [3] describes mathematical and graphical methods for such constructions.

thickness, t, where the length of the anvils is at least equal to w, and their width $b = t$. If $w \gg t$, then as the load, F, is applied there is practically no change in dimension w beneath the anvils, so the deformation zone approximates conditions of plane strain. For now, assume the anvil-workpiece interface is frictionless, thus the maximum shear must occur along lines AEC and DEB since σ_x is a principal stress. Regions DEA and CEB act as two rigid blocks, while regions to the right of AEB and left of DEC are also rigid and move as indicated by the arrows. It is obvious that as the load is increased, the exact pattern shown in Fig. 9-1 must change as the thickness between the anvils diminishes. Suppose, however, that interest lies in calculating the magnitude of load F that will initiate plastic deformation. Then the simple flow field of $DAEBC$ could be viewed as a slip-line field which enables such a prediction. The horizontal stress, σ_y, must be zero, since no restraint forces act in the y direction, whereas the vertical or loading stress, σ_x, must be a principal stress as the surfaces DA and CB are frictionless. For now, assume that σ_z is intermediate between these other stresses in magnitude. From the physical problem it follows that

$$\sigma_x = \sigma_3, \sigma_y = \sigma_1 = 0, \quad \sigma_x < \sigma_z < \sigma_y$$

where $\sigma_z = \sigma_2$ and the sign convention $\sigma_1 > \sigma_2 > \sigma_3$ has been followed. In many cases the use of Mohr's circle provides a useful and clarifying representation of the state of stress; for this example see Fig. 9-2.

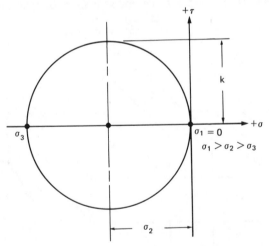

Figure 9-2 Mohr's stress circle related to Fig. 9-1.

The compressive stress required to cause yielding is $\sigma_3 = 2k$ and the corresponding force, F, is $2kwb$. This simple example illustrates to a large extent the basis of slip-line field theory. Knowing the directions of maximum shear and the orientation of principal stresses from the physics of the problem, one can determine loads of concern in terms of the yield shear stress, k. Of course, not many situations would be fully satisfied by a slip-line field composed entirely of straight lines as shown in Fig. 9-1. More complicated fields will be treated later.

9-2 GOVERNING STRESS EQUATIONS

The assumption of plane strain, which is invoked throughout this chapter, implies certain relations. These will be developed so they can be applied directly in the analysis of slip-line fields.

In plane-strain deformation, metal flow is always parallel to a given plane, there being no motion or displacement of metal in a direction perpendicular to the plane of flow. Thus in Fig. 9-1, metal flow occurs in planes parallel to the x-y plane and no movement of metal would occur in the z direction. Considering such an xyz system, and invoking the concept of volume constancy, one can note that

$$d\epsilon_x = -d\epsilon_y, \, d\epsilon_z = 0 \tag{9-1}$$

$$\dot{\epsilon}_x = -\dot{\epsilon}_y, \, \dot{\epsilon}_z = 0 \tag{9-2}$$

$$d\gamma_{xy} \neq 0, \, d\gamma_{yz} = d\gamma_{zx} = 0 \tag{9-3}$$

$$\dot{\gamma}_{xy} \neq 0, \, \dot{\gamma}_{yz} = \dot{\gamma}_{zx} = 0 \tag{9-4}$$

and therefore,

$$\tau_{zy} = \tau_{xz} = 0 \tag{9-5}$$

so σ_z must be a principal stress and $\sigma_z = \sigma_2 = \frac{1}{2}(\sigma_x + \sigma_y)$.[†] Thus, σ_2 must be equivalent to the *mean* or hydrostatic stress at any point in the field of deformation. If the x- and y-axes define principal directions, it follows that

$$\sigma_2 = \frac{\sigma_1 + \sigma_2 + \sigma_3}{3} = \sigma_{mean}$$

and

$$\sigma_1 = \sigma_2 + k, \quad \sigma_3 = \sigma_2 - k \tag{9-6}$$

Thus, one important observation that is always applicable to plane-strain deformation is that the *intermediate stress*, σ_2, is always equal to the mean stress![†] In effect, it is the hydrostatic component of the stress state and has no influence upon yielding as pointed out earlier. Recalling the relation between the Mohr's circles of stress and strain for plastic flow, the position of the mean stress corresponds to the position of the strain circle where $d\epsilon = 0$; this follows from Eqs. (9-1) and (9-5). Also from Eq. (9-1), plane strain is equivalent to distortion in pure shear.

Thus, plane-strain deformation causes a state of stress which may be considered as a pure-shear deformation with a superimposed hydrostatic stress that can vary from one point to another in the region of deformation. See Hill [2] for a discussion. Yielding occurs when the yield shear stress, k, is reached

---

[†]This assumes the von Mises criterion applies. No reference is made to Tresca here.

and it can be shown that the yield criterion for this situation is

$$k^2 = \frac{(\sigma_x - \sigma_y)^2}{4} + \tau_{xy}^2 \qquad (9\text{-}7)$$

Note that $2k = 2Y/\sqrt{3}$ for the von Mises criterion.

In connection with the preceding comments, consider the following general state of stress at a point as shown in Fig. 9-3.

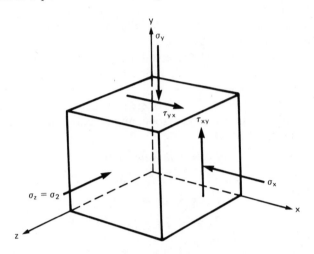

Figure 9-3 Stress element for plane-strain deformation.

Arbitrarily the three normal stresses have been chosen as compressive, with σ_z the intermediate stress, σ_2. This satisfies the conditions of plane strain, and the Mohr's circle for stress and strain rate are shown in Fig. 9-4, where $\sigma_x < \sigma_y$. From Fig. 9-4a, it can be seen that the normal stress acting on the plane upon which k acts is the mean stress σ_2. In terms of incremental strains (or strain rates), this plane is subjected to the maximum shearing strain and zero normal strain; thus, there is no *length* change associated with these planes.

The planes of maximum shear are orthogonal to each other; a network of such orthogonal lines describes a slip-line field. For the general case where such lines are curved, rather than straight, Fig. 9-5 illustrates an elemental section in the field. The shear stress, k, acts along the slip lines, while the mean stress, σ_2, (which equals $-P$, the mean normal pressure) acts perpendicular to these lines and to the face of the element as shown. Compared with Fig. 9-3, the element in Fig. 9-5 is of course rotated by an angle ϕ from the x-axis, which would correspond to an angle 2ϕ on a plot such as Fig. 9-4a. Now as one moves along a slip line, the magnitude of σ_2 can vary from point to point and it is this change that is of greatest significance when this theory is used.

To develop the essential equations, and to employ them correctly, it is necessary to invoke a consistent convention. Families of orthogonal slip lines are usually designated as α- and β-lines and it is essential that they are distin-

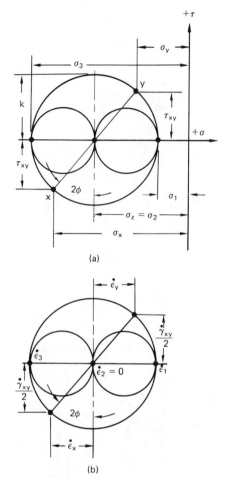

Figure 9-4 (a) Mohr's stress and (b) strain-rate circles for a general state of stress at a point.

guished from each other. The angle of rotation ϕ is considered *positive* for counterclockwise movement from some datum (say the x-axis here) and the *direction* of the largest *algebraic* principal stress, σ_1, lies in the first and third quadrants of the α-β system. Equivalently, the β-lines fall in the first and third quadrants formed by the σ_1 and σ_3 stress axes. Figure 9-6 describes this situation.

Note that a stress of zero is larger than a compressive stress; if all principal stresses are compressive, then the least compressive is the *algebraically* largest. For plane strain, since τ_{yz} and τ_{zx} are zero, the equilibrium equations (see Chap. 1) reduce to:

$$\frac{\partial \sigma_x}{\partial x} + \frac{\partial \tau_{yx}}{\partial y} = 0 = \frac{\partial \sigma_y}{\partial y} + \frac{\partial \tau_{xy}}{\partial x} \qquad (9\text{-}8)$$

and from Fig. 9-4,

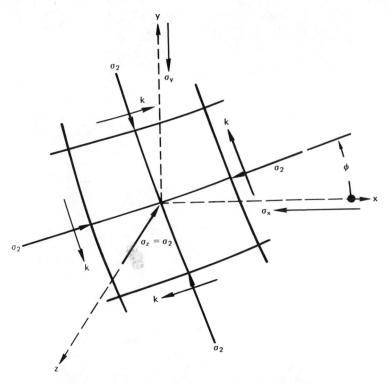

Figure 9-5 Stresses acting on a small curvilinear element.

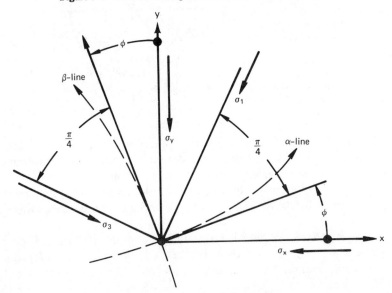

Figure 9-6 Convention for defining α- and β-lines.

$$\sigma_x = \sigma_2 - k \sin 2\phi$$
$$\sigma_y = \sigma_2 + k \sin 2\phi \qquad (9\text{-}9)$$
$$\tau_{xy} = k \cos 2\phi$$

Combining Eqs. (9-8) and (9-9) and then differentiating gives

$$\frac{\partial \sigma_2}{\partial x} - 2k \cos 2\phi \frac{\partial \phi}{\partial x} - 2k \sin 2\phi \frac{\partial \phi}{\partial y} = 0$$

$$\frac{\partial \sigma_2}{\partial y} + 2k \cos 2\phi \frac{\partial \phi}{\partial y} - 2k \sin 2\phi \frac{\partial \phi}{\partial x} = 0 \qquad (9\text{-}10)$$

Equations (9-10) can be applied to any set of arbitrary axes x', y' in the x-y plane by measuring ϕ from x' to the α-line. By orienting x' and y' tangent to the α- and β-lines at point O, $\phi = 0$. Then Eqs. (9-10) reduce to functions of one variable and partial derivatives are unnecessary. Thus,

$$\frac{d\sigma_2}{dx'} - 2k \frac{d\phi}{dx'} = 0$$

$$\frac{d\sigma_2}{dy'} + 2k \frac{d\phi}{dy'} = 0 \qquad (9\text{-}11)$$

which upon integrating gives

$$\sigma_2 - 2k\phi = C_1 = \text{constant along an } \alpha\text{-line}$$

$$\sigma_2 + 2k\phi = C_2 = \text{constant along a } \beta\text{-line} \qquad (9\text{-}12)$$

Physically, movement along an α- or β-line causes σ_2 to change by

$$\Delta\sigma_2 = 2k\,\Delta\phi_\alpha \qquad \text{on an } \alpha\text{-line}$$

or

$$\Delta\sigma_2 = -2k\,\Delta\phi_\beta \qquad \text{on a } \beta\text{-line} \qquad (9\text{-}13)$$

If σ_2 is replaced by $-P$ (the normal pressure on a slip line), Eqs. (9-12) are known as the Hencky equations and are generally written as

$$P + 2k\phi = C_1 \ (\alpha\text{-line}) \quad \text{or} \quad \Delta P = -2k\Delta\phi_\alpha$$

$$P - 2k\phi = C_2 \ (\beta\text{-line}) \quad \text{or} \quad \Delta P = 2k\Delta\phi_\beta \qquad (9\text{-}14)$$

In general, the values of C_1 and C_2 vary from one slip line to another, but numerical evaluation of these constants is generally unnecessary. In effect, Eqs. (9-12) or (9-14) are the equilibrium equations expressed along a slip line, and the inclusion of equilibrium considerations is one of the major differences between slip-line field theory and the upper-bound approach.

9-3 BOUNDARY CONDITIONS

Regardless of the type of slip-line field posed for a particular problem, the establishment of one principal stress is usually determined at some boundary. Among the most useful are the following:

a) *Free Surface* (plastic zone extends beyond tool). Since there are no normal or tangential stresses at a free surface, the α- and β-lines must meet such a surface at 45°.

b) *Frictionless Interface* (between material and tool). As with (a), the α- and β-lines meet the surface at 45°.

c) *Sticking Friction at Interface*. Wherever the slip lines meet an interface with sticking friction, then k is reached at the interface so one slip line, say α, meets the interface tangentially while the other, β, meets it at right angles.

Equation (9-12) can be used to establish an additional restriction on the shape of a statically admissible field. Consider the region bounded by the two α- and β-lines in Fig. 9-7. The difference between σ_2 at A and C can be established by traversing two different paths. First consider the path A to D to C. In moving on an α-line from A to D,

$$\sigma_{2D} = \sigma_{2A} + 2k(\phi_D - \phi_A) \tag{9-15a}$$

In moving along a β-line from D to C,

$$\sigma_{2C} = \sigma_{2D} - 2k(\phi_C - \phi_D) \tag{9-15b}$$

so, combining Eqs. (9-15a) and (9-15b) gives

$$\sigma_{2C} - \sigma_{2A} = 2k(2\phi_D - \phi_A - \phi_C) \tag{9-16}$$

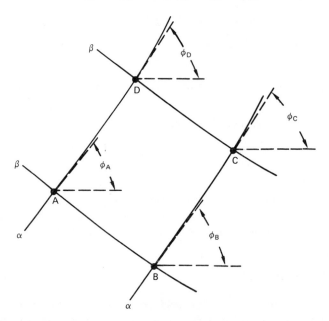

Figure 9-7 Two pairs of α- and β-lines for analyzing the change in the mean normal stress by traversing two different paths.

Alternatively, consider the path A to B to C. In moving on a β-line from A to B,

$$\sigma_{2B} = \sigma_{2A} - 2k(\phi_B - \phi_A) \tag{9-17a}$$

then from B to C on an α-line gives

$$\sigma_{2C} = \sigma_{2B} + 2k(\phi_C - \phi_B) \tag{9-17b}$$

so

$$\sigma_{2C} - \sigma_{2A} = -2k(2\phi_B - \phi_A - \phi_C) \tag{9-17c}$$

Equating Eqs. (9-17c) and (9-16) gives

$$\phi_A - \phi_B = \phi_D - \phi_C$$

or

$$\phi_A - \phi_D = \phi_B - \phi_C \tag{9-18}$$

Equation (9-18) implies that a net of α- and β-lines must be so constructed that the *change* of ϕ is the same along a given family of lines moving from one intersection with the opposite family to another such intersection. This is in addition to the orthogonality requirement, and indicates that the angular change, rather than the actual length traversed, is of sole importance.

9-4 PLANE-STRAIN INDENTATION

There are two particularly simple fields that meet these requirements. One is a net of straight lines as shown in Fig. 9-8a. Since there is no curvature, the stress σ_2 remains the same throughout the field. Another simple field is the centered fan of Fig. 9-8b. There the value of σ_2 is the same everywhere along a radial line, but changes from one radial line to another. At the singular point, O, σ_2 is indeterminate.

A number of problems can be solved by combining these two simple fields. In doing this one should always realize that the normal to a free surface or a frictionless surface is a direction of principal stress; therefore, the slip lines must meet these surfaces at 45°. Figure 9-9 shows how these can be com-

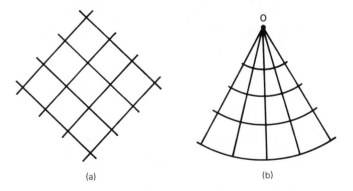

(a)   (b)

Figure 9-8   (a) A net of straight lines and (b) a centered fan.

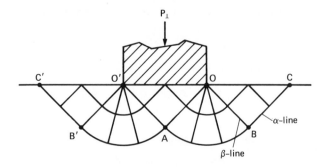

Figure 9-9 A slip-line field for plane-strain indentation, of a semi-infinite slab.

bined to form a slip-line field† for an indentation problem, while Fig. 9-10a shows the right-hand half of this field with the changing stress state as indicated.

Along OC, $\sigma_y = \sigma_1 = 0$, $\sigma_x = \sigma_3$ and is compressive, and $\sigma_z = \sigma_2 = \frac{1}{2}\sigma_3$ as shown in Fig. 9-10b. Applying the convention in Fig. 9-6, $CBAO'$ is an α-line, while radial lines such as OB and OA are β-lines. Since triangle OBC is, in effect, a net of straight lines, Eq. (9-6) gives

$$\sigma_1 = 0, \sigma_2 = -k, \sigma_3 = -2k$$

everywhere in OBC. In moving clockwise from B to A along an α-line, through

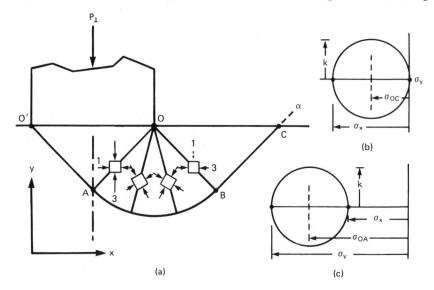

(a)   (c)

Figure 9-10 A more detailed illustration of Fig. 9-9 showing (a) the changing stress state, (b) Mohr's stress circle for triangle OBC, and (c) Mohr's stress circle for triangle AOO'.

†This field is due to L. Prandtl, *Nachr. Ges. Wiss. Göttingen*, 74 (1920). Another field, which gives the same answer, was proposed by R. Hill [2].

an angle $\Delta\phi_\alpha = -\pi/2$, Eq. (9-12) gives

$$\sigma_{2A} = \sigma_{2B} + 2k\Delta\phi_\alpha$$

Since $\sigma_{2B} = -k$ and $\Delta\phi_\alpha = -\pi/2$,

$$\sigma_{2A} = -k + 2k\left(\frac{-\pi}{2}\right) = -k(1 + \pi).$$

This is also the value of σ_2 throughout AOO', since this region is composed of straight lines only and is acted upon by compressive stresses in all principal directions; this is shown in Fig. 9-10c.

To determine the tool pressure $P_\perp$, which is $-\sigma_y = -\sigma_3$,

$$\sigma_{300'} = \sigma_{2A} - k = -k(1 + \pi) - k = -2k\left(1 + \frac{\pi}{2}\right)$$

thus,

$$P_\perp = -\sigma_{300'} = 2k\left(1 + \frac{\pi}{2}\right)$$

so

$$P_\perp/2k = 2.57 \tag{9-19}$$

If the von Mises criterion is used, $2k = 1.15Y$ so $P_\perp = 2.97Y$. As this is analogous to a two-dimensional hardness test, $P_\perp$ reflects the hardness; often, the hardness (using proper units) is quoted as being equal to $3Y$ and this problem indicates the basis for such a relationship.

It is also of interest to show that a yield criterion has not been violated anywhere in the field. Within the triangle AOO', $\sigma_2 = -k(1 + \pi)$, $\sigma_1 = -\pi k$, and $\sigma_3 = -2k(1 + \pi/2)$. A substitution for σ_1 and σ_3 gives $\sigma_1 - \sigma_3 = 2k$, so both the Tresca and von Mises yield criteria are satisfied. Similar findings result if other sections of this field are investigated.

Reviewing this problem indicates the following:

1. Regions OBC and OAO' experience constant but different values of σ_2.
2. The sector OAB provides gradual deformation of metal into OBC and σ_2 increases from OBC to OAO'.
3. If the left-hand half of the field had been analyzed, the mirror image of line ABC would be a β-line while the counterpart of OB is an α-line. Repeating this problem again gives $P_\perp/2k = 2.57$ and it is suggested that the reader prove this, since it emphasizes why the convention for labeling α- and β-lines in Fig. 9-6 is essential.

9-5 VELOCITIES IN A SLIP-LINE FIELD†

In the pressure calculation above, no use was made of a hodograph to calculate loads, as was required in the upper-bound approach. The construction of hodographs in slip-line analyses are necessary, however, in connection with the

---

†Velocity equations attributed to Geiringer [5] are not used in this text, since hodographs are used whenever velocities are of concern.

following:

1. To assure that the field is kinematically admissible.
2. To determine the relative percentages of energy dissipated by gradual deformation as opposed to intense shear.
3. To predict the approximate change in the shape of initially straight lines as they proceed through the deformation zone to the exit.

Before constructing the hodograph for the indentation problem, several general comments about velocities of particles in a slip-line field may prove helpful. These are:

1. Within any individual zone of constant σ_2, the absolute velocity is constant.
2. In entering or leaving a region where σ_2 will gradually change, a jump discontinuity may or may not occur.
3. The velocity discontinuity along a single slip line has constant magnitude; that is, there is no normal strain. Along straight slip lines, such as a radial line in a centered fan, both the magnitude and direction of the velocity discontiniuty are constant.
4. As a particle moves through a region of changing σ_2, its absolute velocity changes in direction and, in many cases, magnitude.
5. Shear always occurs at the boundary between the field and undeformed metal outside the field, and most of the energy dissipation occurs at those locations.

Figure 9-11a shows the previous indentation example and Fig. 9-11b is the hodograph for the right half of this field. Region OAD moves downward at

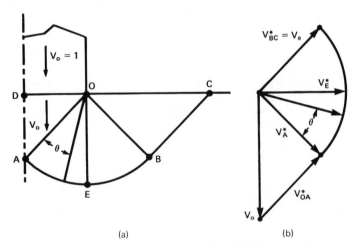

(a)   (b)

Figure 9-11   (a) A partial field and (b) associated hodograph for Fig. 9-10.

the punch speed, V_o. At AO there is a jump discontinuity V^*_{OA}, parallel to AO. A particle just inside OAB at point A must have an absolute velocity that is tangent to AEB at point A; this is drawn as V^*_A on the hodograph, implying that shear occurs all along AEB. Since the magnitude of the velocity discontinuity is constant *along* a slip line, the magnitude of V^*_A is constant along AEB. Using O as center, an arc of radius V^*_A is then laid off on the hodograph as shown; since AOB on the field is 90°, this arc on the hodograph includes a right angle. Note that V^*_E has the same magnitude as V^*_A, but its direction reflects a line tangent to E (or perpendicular to OE) on AEB. The changing velocity (here direction only) through region AOB is gradual, and across OB there is *no abrupt* discontinuity. That is, the velocity everywhere in triangle OBC is V^*_B, which is parallel to BC. To summarize the above it should be understood that,

1. Intense shear occurs along AO (velocity is V^*_{OA}), AEB (velocity $V^*_A = V^*_E = V^*_B$ etc.), and along BC (exit velocity $V_e = V^*_B$). Here, velocity *magnitudes* are implied.
2. Energy is also dissipated during the gradual deformation occurring in AOB.
3. For conservation of mass, which satisfies incompressibility or constant volume, the flow rate across OD must equal the flow rate across OC.

9-6 PLANE-STRAIN EXTRUSION

Consider the extrusion problem completed earlier using an upper-bound analysis, Sec. 8-4, where $r = 0.5$ and $\alpha = 30°$. A slip-line solution could make use of the field shown in Fig. 9-12a, the top half of which is magnified in Fig. 9-12b. Along OC, x and y are the 1 and 3 directions. Figure 9-12c is the Mohr's circle for stresses along OC.

The stress $\sigma_1 = \sigma_x = 0$ at the exit section, so the α- and β-lines are determined as shown. Now $\sigma_{2OC} = -k$ everywhere along OC, thus $\sigma_{2C} = -k$. Therefore, with BC as an α-line,

$$\sigma_{2B} = \sigma_{2C} + 2k(\phi_B - \phi_C)$$

and since

$$\sigma_{2C} = -k \quad \text{and} \quad (\phi_B - \phi_C) = \frac{-\pi}{6} \quad \text{(clockwise)},$$

$$\sigma_{2B} = -k\left(1 + \frac{\pi}{3}\right)$$

The Mohr's circle for the stress state along OB, Fig. 9-12d, indicates

$$\sigma_{3B} = \sigma_{3AOB} = -k\left(2 + \frac{\pi}{3}\right)$$

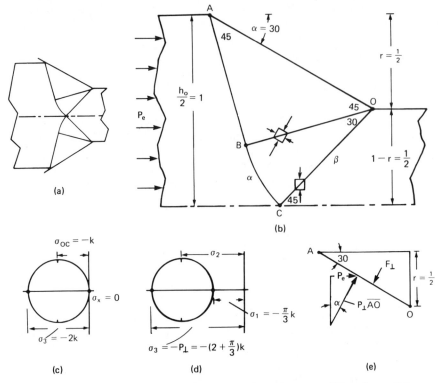

Figure 9-12 Plane-strain, frictionless extrusion where $\alpha = 30°$ and $r = 0.5$ showing (a) a proposed field, (b) an enlargement of one half the field, (c) Mohr's stress circle along OC, (d) Mohr's stress circle along OB, and (e) essentials for calculating the extrusion pressure from a force balance.

Thus, the pressure perpendicular to the die along AO is $-\sigma_{3B}$, or

$$P_\perp = k\left(2 + \frac{\pi}{3}\right) \qquad (9\text{-}20)$$

Figure 9-12e is now used to determine the extrusion pressure, P_e, by taking a force balance. The sum of the forces in the x direction must be zero. Since the force perpendicular to the die wall is

$$F_\perp = P_\perp(\overline{AO}) = P_\perp \frac{r}{\sin \alpha}$$

its horizontal component is

$$F_x = F_\perp \sin \alpha = rP_\perp$$

Equating the force acting on the billet due to P_e with F_x gives

$$P_e\left(\frac{h_o}{2}\right) = F_x = rP_\perp$$

182

such that

$$P_e = \frac{2r}{h_o}(k)(2 + \pi/3)$$

or $\qquad \frac{P_e}{2k} \approx 0.762 \quad \text{since} \quad \frac{h_o}{2} = 1 \quad \text{and} \quad r = \frac{1}{2} \qquad (9\text{-}21)$

Recall that one upper-bound solution for this problem gave a value of 0.78 for $P_e/2k$.

The example just considered is one of a series of cases for which the appropriate slip-line field is composed of a constant-stress field and a centered fan. The general geometry of this type of field is shown in Fig. 9-13 where, for convenience, the semi-thickness of the slab, h_o, is taken as unity. Then the

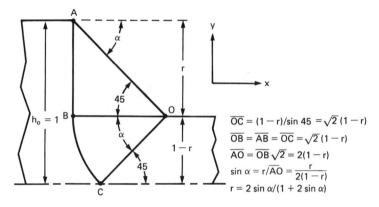

$$\overline{OC} = (1 - r)/\sin 45 = \sqrt{2}\,(1 - r)$$
$$\overline{OB} = \overline{AB} = \overline{OC} = \sqrt{2}\,(1 - r)$$
$$\overline{AO} = \overline{OB}\,\sqrt{2} = 2(1 - r)$$
$$\sin \alpha = r/\overline{AO} = \frac{r}{2(1 - r)}$$
$$r = 2\sin\alpha/(1 + 2\sin\alpha)$$

Figure 9-13 The general geometry for an acceptable field used in plane-strain extrusion or drawing where the reduction, r, and semi-die angle, α, are related by $r = 2\sin\alpha/(1 + 2\sin\alpha)$.

exiting semi-thickness, h_f, becomes equal to $1 - r$, where r is the reduction, $(h_o - h_f)/h_o$. As shown in Fig. 9-13, this form of field is appropriate when the reduction and die angle are related by

$$\sin \alpha = \frac{r}{2(1 - r)}$$

or, $\qquad\qquad r = \frac{2\sin\alpha}{(1 + 2\sin\alpha)} \qquad (9\text{-}22)$

If the relation in Eq. (9-22) is violated, this simple field is not appropriate. For situations where Eq. (9-22) is satisfied, the extrusion pressure is found using the same procedures just followed, and the general expression for the die pressure is

$$P_\perp = 2k(1 + \alpha) \qquad (9\text{-}23)$$

while the normalized extrusion pressure is

$$\frac{P_e}{2k} = r(1 + \alpha) = w_a/2k \qquad (9\text{-}24)$$

It is of interest to compare values of P_e found with slip-line analyses to the values based upon ideal homogeneous deformation. The latter gives

$$P_e = \int \bar{\sigma} \, d\bar{\epsilon} = 2k \ln \left(\frac{1}{1-r}\right) = w_i \qquad (9\text{-}25)$$

Defining the process efficiency, η, as the ratio of these predictions, w_i/w_a, gives

$$\eta = \frac{2k \ln \left[1/(1-r)\right]}{2kr(1+\alpha)} = \frac{\ln \left[1/(1-r)\right]}{r(1+\alpha)} \qquad (9\text{-}26)$$

If, in the previous example, the reduction had occurred by drawing instead of extrusion, a similar analysis would be used; refer to Fig. 9-13. The outlet drawing stress, σ_d, would be tensile; this would be σ_1 in the usual notation. OC is thus established as a β-line, while CBA is an α-line. Now,

$$\sigma_{2OC} = \sigma_d - k = \sigma_{2C}$$

$$\sigma_{2B} = \sigma_{2C} + 2k(\phi_C - \phi_B) = \sigma_{2C} - 2k\alpha = \sigma_d - k(1 + 2\alpha)$$

In AOB,
$$\sigma_3 = \sigma_{2B} - k$$

thus,
$$P_\perp = -\sigma_d + 2k(1 + \alpha)$$

Equating forces in a manner similar to Fig. 9-12e gives

$$F_x = rP_\perp = \sigma_d(1 - r)$$

Thus,
$$r[2k(1 + \alpha) - \sigma_d] = \sigma_d(1 - r)$$

so
$$\frac{\sigma_d}{2k} = r(1 + \alpha) \qquad (9\text{-}27)$$

which is identical in form to Eq. (9-24).

9-7 ENERGY DISSIPATION IN A SLIP-LINE FIELD

To illustrate the point of concern, consider the extrusion conditions of Fig. 9-12b. Intense shear, due to abrupt velocity discontinuities, occurs along AB, BC, and CO. The hodograph for this condition is given in Fig. 9-14b. By either a graphical approach or trigonometry, the following may be determined:

$$V_{AB}^* = \frac{V_o}{\sqrt{2}} = \frac{1}{\sqrt{2}} \text{ for } V_o = \text{unity}, \quad V_{BC}^* = V_{OC}^* = \frac{1}{\sqrt{2}}$$

The lengths along which these jumps occur are

$$\overline{AB} = \frac{1}{\sqrt{2}}, \quad \overline{BC} = \frac{\pi}{6}\left(\frac{1}{\sqrt{2}}\right), \text{ and } \overline{OC} = \frac{1}{\sqrt{2}}$$

Taking the usual summation, the energy dissipation on lines of intense shear is

$$k(\overline{AB} \cdot V_{AB}^* + \overline{BC} \cdot V_{BC}^* + \overline{OC} \cdot V_{OC}^*) = 1.262k$$

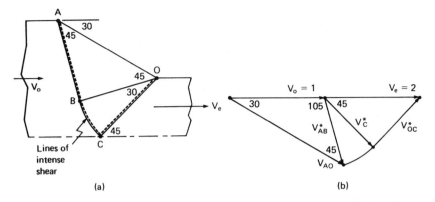

Figure 9-14 (a) Illustration of regions of intense shear in a slip-line field for extrusion and (b) corresponding hodograph.

Earlier it was determined that $P_e/2k = 0.762$ for this problem, thus $P_e = 1.542k$. In effect, the percent of the energy dissipated along lines of intense shear is $1.262/1.524 = 83\%$. The remaining 17% is expended during the gradual deformation of metal in region OBC. It is to be realized that this is not an upper-bound analysis; rather, it indicates how the energy dissipated because of velocity discontinuities may be found with a slip-line field.

9-8 METAL DISTORTION

Once a satisfactory slip-line field is found to a problem, the corresponding hodograph can be drawn to check whether the field is self-consistent. The hodograph and the field can be used to determine the distortion of the metal formed.

As an example, consider a $2 : 1$ reduction in a $90°$ die. The appropriate field, shown in Fig. 9-15a, consists of a single centered fan; the hodograph is shown in Fig. 9-15b. In the triangle to the right of OA, the material forms a dead-metal zone. A particle entering the field at A suffers a velocity discontinuity V_A^* parallel to the tangent of the arc at A; its absolute velocity is parallel to OA. A particle entering at C suffers a discontinuity, V_C^*, perpendicular to OC. The magnitudes of V_A^* and V_C^* are equal, as is every other discontinuity along arc AG; they differ only in direction. All particles on the same radial line have the same velocity, which changes as they move through the field. By the time they reach OG, their absolute velocity is V_G. Here they suffer a discontinuity V_{OG}^*, on crossing OG, and acquire the final velocity, V_f, which is $2V_o$.

Stream lines can be drawn for any particle. Consider a particle on line 3 of Fig. 9-15c. As it enters the field mid-way between C and D, it acquires

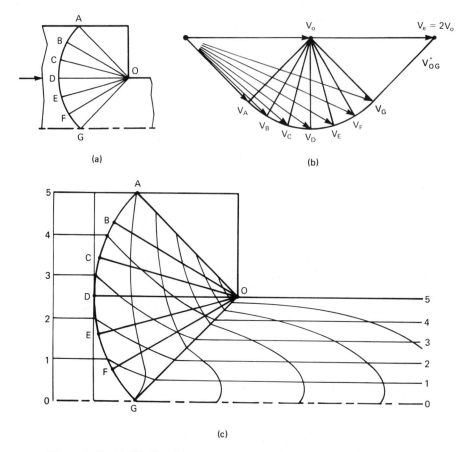

Figure 9-15 (a) Slip-line field, (b) hodograph, and (c) predicted distortion of vertical grid lines when extruding through a 90° die.

an absolute velocity mid-way between V_C and V_D. Hence, the new direction of travel can be drawn at this point. This direction gradually changes, however, and by the time the particle crosses radius D, its path must be parallel to V_D. As it approaches OG, its direction must be parallel to V_G; then it is abruptly sheared so that its path is parallel to V_f. If this construction is made correctly, the particle should emerge from the field at a height equal to $\frac{3}{5}$ the final half-thickness, just as it entered at a height $\frac{3}{5}$ of the initial half-thickness. Paths of other particles on lines 0 through 5 are also shown in Fig. 9-15c.

The distortion in the field is found by considering the velocity magnitudes along the various paths. Consider a vertical line at time zero. For any increment of time, Δt, the distance traveled, Δs, will be proportional to $V\Delta t$, where V is the velocity characteristic of its position in the field. By following particles on paths 0 through 5, the distortion of a vertical grid line would end up as shown.

The most severe distortion occurs near the surface, because particles entering at the top of the field move much slower than those lower in the field.

9-9 INDENTATION AS A FUNCTION OF WORKPIECE GEOMETRY

Consider the indentation of a large slab of thickness, H, by two opposing flat indentors of width, L, as shown in Fig. 9-16. For the special case where $L = H$, the simple slip-line field of Fig. 9-17 is appropriate. Along line AO, $\sigma_x = 0$, so $\sigma_2 = -k$. This condition prevails throughout triangle AOO'. Therefore,

$$\sigma_y = \sigma_2 - k = -2k, \quad P_\perp = -\sigma_y$$

$$\frac{P_\perp}{2k} = 1 \tag{9-28}$$

As H/L becomes larger than one, a different field must be used to satisfy field requirements. This is illustrated in Fig. 9-18 for $H/L = 5.43$, which is a net having angular increments of $15°$. In triangle AOO',

$$\sigma_y = -P_\perp, \quad \sigma_{2A} = \sigma_y + k = -P_\perp + k$$

Moving along an α-line to 0, 1

$$\sigma_{2(0,1)} = \sigma_{2A} + 2k\Delta\phi_\alpha$$

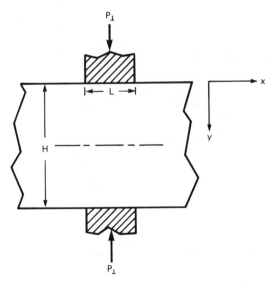

Figure 9-16   General geometry of slab indentation.

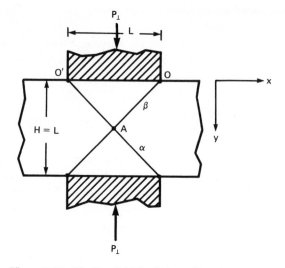

Figure 9-17 Slip-line field for indentation when $H/L = 1$.

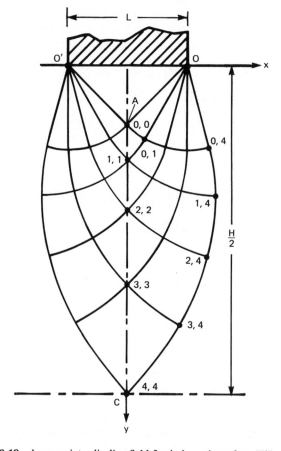

Figure 9-18 Appropriate slip-line field for indentation when $H/L = 5.43$.

and in moving back along a β-line to $(1,1)$,

$$\sigma_{2(1,1)} = \sigma_{2(0,1)} - 2k\Delta\phi_\beta$$
$$= \sigma_{2A} + 2k(\Delta\phi_\alpha - \Delta\phi_\beta)$$
$$= -P_\perp + k + 2k(\Delta\phi_\alpha - \Delta\phi_\beta)$$

Now,

$$\sigma_{x(1,1)} = \sigma_{2(1,1)} + k$$

so that

$$\sigma_{x(1,1)} = -P_\perp + 2k + 2k(\Delta\phi_\alpha - \Delta\phi_\beta)$$

In this case $\Delta\phi_\alpha = +\pi/12$ and $\Delta\phi_\beta = -\pi/12$. At any point (n, n) along the centerline, $\Delta\phi_\alpha = n\pi/12$, $\Delta\phi_\beta = -n\pi/12$, so

$$\sigma_{x(n,n)} = -P_\perp + 2k(1 + \Delta\phi_\alpha - \Delta\phi_\beta) = -P_\perp + 2k(1 + 2\Delta\phi) \quad (9\text{-}29)$$

If the slab is unconstrained in the x direction, an appropriate boundary condition is $F_x = 0$. The net force F_x in the x direction is equal to the integral of σ_x along the centerline.

$$F_x = 0 = \int_0^{h/2} \sigma_x \, dy \quad (9\text{-}30)$$

Substituting Eq. (9-29) into (9-30),

$$\int_0^{h/2} [-P_\perp + 2k(1 + 2\Delta\phi)] \, dy = 0$$

$$P_\perp \cdot \frac{h}{2} = 2k\left(\frac{h}{2}\right) + 2k \int_0^{h/2} 2\Delta\phi \, dy$$

$$P_\perp = 2k + \frac{4k}{h} \int_0^{h/2} 2\Delta\phi \, dy \quad (9\text{-}31)$$

Values of $\Delta\phi$ as a function of y are given in Fig. 9-19 for a 15° net. For those readers interested in more detailed calculations, comparable values for a 5° net are given in the Appendix at the end of this chapter. Using the results in Fig. 9-19, the integration of Eq. (9-31) can be performed graphically by first plotting $2\Delta\phi$ versus y, or numerically by use of the trapezoidal rule.

The results of such calculations are summarized in Fig. 9-20, where $P_\perp/2k$ is plotted versus H/L and acceptable slip-line fields are shown for various ratios. At values of $H/L > 8.75$, $P_\perp/2k$ for this field exceeds $(1 + \pi/2)$, which is the value for the non-penetrating indentation illustrated in Fig. 9-9. Physically, this means that non-penetrating deformation should be expected when $H/L > 8.75$ and penetrating deformation for smaller values of H/L.†

A useful conclusion is that for valid hardness testing, the indentation should be non-penetrating, and the thickness of the metal should be four to five times the size of the indentation impression. (Theoretically, $H/L > 8.75 =$

---

†Hill [2] provided the first correct plot of Fig. 9-20 while Green [6] discussed the frictionless case when $H/L < 1$.

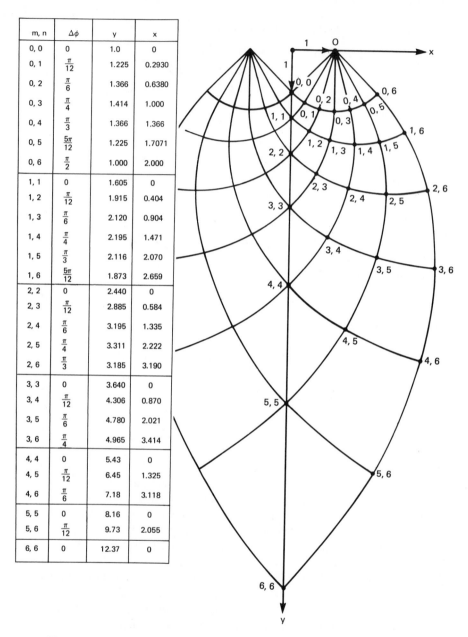

m, n	$\Delta\phi$	y	x
0, 0	0	1.0	0
0, 1	$\frac{\pi}{12}$	1.225	0.2930
0, 2	$\frac{\pi}{6}$	1.366	0.6380
0, 3	$\frac{\pi}{4}$	1.414	1.000
0, 4	$\frac{\pi}{3}$	1.366	1.366
0, 5	$\frac{5\pi}{12}$	1.225	1.7071
0, 6	$\frac{\pi}{2}$	1.000	2.000
1, 1	0	1.605	0
1, 2	$\frac{\pi}{12}$	1.915	0.404
1, 3	$\frac{\pi}{6}$	2.120	0.904
1, 4	$\frac{\pi}{4}$	2.195	1.471
1, 5	$\frac{\pi}{3}$	2.116	2.070
1, 6	$\frac{5\pi}{12}$	1.873	2.659
2, 2	0	2.440	0
2, 3	$\frac{\pi}{12}$	2.885	0.584
2, 4	$\frac{\pi}{6}$	3.195	1.335
2, 5	$\frac{\pi}{4}$	3.311	2.222
2, 6	$\frac{\pi}{3}$	3.185	3.190
3, 3	0	3.640	0
3, 4	$\frac{\pi}{12}$	4.306	0.870
3, 5	$\frac{\pi}{6}$	4.780	2.021
3, 6	$\frac{\pi}{4}$	4.965	3.414
4, 4	0	5.43	0
4, 5	$\frac{\pi}{12}$	6.45	1.325
4, 6	$\frac{\pi}{6}$	7.18	3.118
5, 5	0	8.16	0
5, 6	$\frac{\pi}{12}$	9.73	2.055
6, 6	0	12.37	0

Figure 9-19 Slip-line field for two centered fans showing x and y as functions of $\Delta\phi$ using a 15° net.

4.37 for plane-strain indentation on a flat frictionless substrate.) Note that although either type of field may fit the problem for a range of H/L values, the one giving the *lowest* value of $P_\perp/2k$ is appropriate, since any field indicating

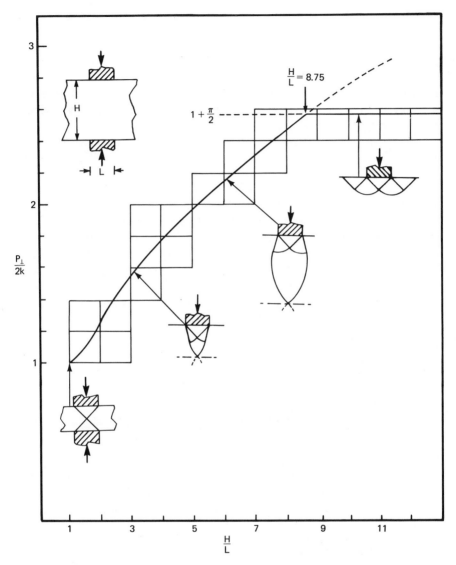

Figure 9-20 Normalized indentation pressure versus H/L where appropriate fields are shown for several values of H/L. This concept was first proposed by Hill [2].

values of $P_\perp/2k$ higher than this will violate equilibrium outside the predicted region of deformation.

The solution of this problem also has important implications in metal forming. Sheet rolling and slab drawing with low-angle dies are similar to indentation. Examination of such fields shows that the inhomogeneity of deformation increases with H/L, and the variation of σ_x from centerline to surface suggests that residual stresses in the x direction should also increase with H/L. For relatively homogeneous deformation, H/L should be kept as low as possible, although frictional constraint and high roll pressures will result with low H/L. Too often in laboratory experiments, investigators have used low reductions per pass in mistaken attempts to achieve homogeneous rolling deformation. However, low reductions cause low values of roll contact L and high H/L, especially in the early stages of rolling when H is high, so a high degree of inhomogeneity is produced.

The field in Fig. 9-19 can be used to analyze drawing and extrusion operations with large H/L values, where the reductions are lower than those whose fields are described by Eq. (9-22) (i.e., $r < 2 \sin \alpha/(1 + 2 \sin \alpha)$).

Figure 9-21a shows an appropriate portion of this field fitted to a frictionless, plane-strain extrusion where $\alpha = 15°$ and $r = 0.111$. Only the portion of the field bounded by α- and β-lines that cross the centerline at $45°$ is used. The appropriate boundary condition is $F_x = \int \sigma_x dy + \int \tau_{xy} dx = 0$. Everywhere along a cut through A, $(0, 1)$, $(1, 2) \ldots (n, n + 1)$ the α- and β-lines are $45°$ to the x-axis, so $\sigma_x = \sigma_1$ is a principal stress; $\tau_{xy} = 0$, so $F_x = \int \sigma_x dy$. The values of the coordinates of points A, $(0, 1) \ldots (n, n + 1)$ in Fig. 9-21 are now designated as x' and y'. These may be transformed to a new x, y coordinate system, with the origin at A, according to

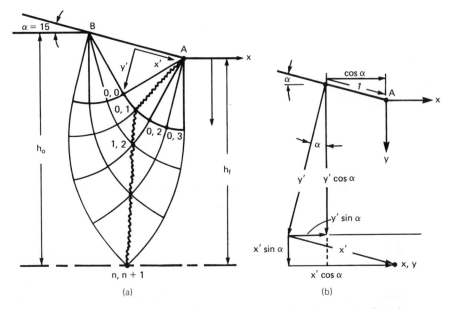

Figure 9-21   (a) One half of an appropriate field for drawing or extrusion when $r < 2 \sin \alpha/(1 + 2 \sin \alpha)$ and (b) a transformed-coordinate system. Plane-strain, frictionless conditions are implied. Along the cut A, $(0, 1)$, $(n, n + 1)$, $\sigma_x = \sigma_1$ since the α- and β-lines are at $45°$ to x.

$$x = (x' - 1) \cos \alpha - y' \sin \alpha \qquad (9\text{-}32a)$$

and $\qquad\qquad y = (x' - 1) \sin \alpha + y' \cos \alpha \qquad (9\text{-}32b)$

as related to Fig. 9-21b. The stress state at any point $(n, n + 1)$ may then be found in terms of $P_\perp$. In AOB and along line AO, $P_{AB} = P_\perp - k$. Rotating on an α-line through $\Delta\phi = +\pi/12$ to $(0, 1)$ gives

$$P_{0,1} = P_{OA} - 2k\left(\frac{\pi}{12}\right) = P_\perp - k - 2k\left(\frac{\pi}{12}\right) \qquad (9\text{-}33)$$

Movement to any point $(n, n + 1)$ requires an additional rotation of $+n(\pi/12)$ on an α-line and $-n(\pi/12)$ on a β-line, so that

$$P_{n,n+1} = P_\perp - k - \frac{2k(2n + 1)\pi}{12} \qquad (9\text{-}34)$$

and $\qquad\qquad \sigma_{x(n,n+1)} = -P_\perp + 2k\left[1 + \frac{(2n + 1)\pi}{12}\right] \qquad (9\text{-}35)$

The boundary condition is

$$F_x = \int_o^{h_f} \sigma_x \, dy = 0 \qquad (9\text{-}36)$$

where h_f is the y value of the centerline on Fig. 9-21a. Thus,

$$F_x = 0 = \int_o^{h_f} \left(-P_\perp + 2k\left[1 + \frac{(2n + 1)\pi}{12}\right]\right) dy \qquad (9\text{-}37)$$

Since $P_\perp$ and k are independent of y,

$$\frac{P_\perp}{2k} = 1 + \frac{\pi}{12h_f} \int_o^{h_f} (2n + 1) \, dy \qquad (9\text{-}38)$$

This may be evaluated graphically or numerically for any value of h_f. Once $P_\perp$ is found, a force balance on the die wall may be used to determine the extrusion pressure, P_e, as

$$P_e = \frac{2P_\perp \sin \alpha}{h_f + 2 \sin \alpha} \qquad (9\text{-}39)$$

The mechanical efficiency for such frictionless extrusions may be found by comparing these $P_e/2k$ values with those for homogeneous work, and such results for $\alpha = 15°$ are shown in Fig. 9-22. It is clear that efficiency decreases sharply at low reductions (i.e., high H/L values) because of redundant work. Note that if rod drawing was done in place of extrusion, the same values of drawing stress, σ_d, would be found in place of P_e for the same values of reduction, r, and semi-die angle, α.

It should be pointed out that if α is large enough in combination with low enough values of r, an alternative field for non-penetrating reduction is possible; see Fig. 9-23a. This is analogous to one half a hardness test using a wedge-shaped indenter. The necessary condition for this mode of deformation

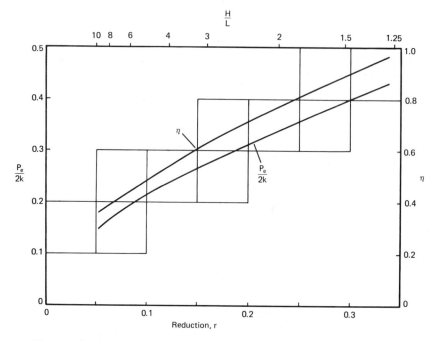

Figure 9-22 Variation in deformation efficiency and normalized extrusion pressure for plane-strain, frictionless extrusion using a semi-die angle of 15°.

is that the die pressure reaches a level such that

$$P_\perp = 2k\left(1 + \frac{\pi}{2} - \alpha\right) \tag{9-40}$$

Material will pile up on the inlet side of the die causing the contact length between the die and metal to increase, as shown in Fig. 9-23b, until this length becomes large enough to lead to penetration. This is often called bulging.[†] Figure 9-24 illustrates how the drawing stress changes with reduction and die angle and indicates the smallest reduction possible before bulging is expected.

Figure 9-25 shows how the hydrostatic pressure ($P = -\sigma_2$) may vary in the deformation zone for strip drawing using the combination of α and r indicated. The key point is that the mean stress, σ_2, may become *tensile* near the centerline (shown by negative values). This hydrostatic tension increases as α increases and r decreases, as shown on Fig. 9-26. As a consequence, structural damage, in the form of void formation may result in the drawn strip.

Slip-line fields can also be used to solve problems involving sticking friction. Let us consider compression of slabs between rough indentors, where sticking friction prevails. This is a condition that approximates many examples

†See Hill [2].

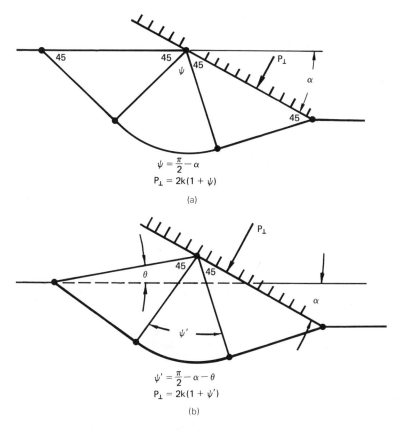

$$\psi = \frac{\pi}{2} - \alpha$$
$$P_\perp = 2k(1 + \psi)$$

(a)

$$\psi' = \frac{\pi}{2} - \alpha - \theta$$
$$P_\perp = 2k(1 + \psi')$$

(b)

Figure 9-23 (a) An alternate field with a combination of large α and small r to start bulging and (b) conditions that prevail during bulging. See Hill [2].

of hot forging. With sticking friction, the slip lines must be *parallel to* or perpendicular to the surfaces, where the slip lines meet the surface, but they need not meet the tool surface everywhere, as a "dead-metal zone" or "cap" can form beneath the indentors. An appropriate slip-line field, taken from reference [1], for a half-slab is given in Fig. 9-27.

How much of this field should be used in a given problem depends upon H/L. The technique for finding the average pressure, $(P_\perp)_a$, is similar to that just used for indentation. Here the appropriate boundary condition is $\sigma_x = 0$ along the left-hand side of the field, and values of $P_\perp = -\sigma_y$ can be found along the centerline. $F_y = \int \sigma_y \, dx$ can then be determined by graphical or numerical integration and $(P_\perp)_a = F_y/L$. Results of such calculations are shown graphically in Fig. 9-28, where $(P_\perp/2k)_a$ is plotted as a function of L/H. The slab solution of $(P_\perp/2k)_a = 1 + L/4H$, given by Eq. (7-25), is also shown for comparison.

195

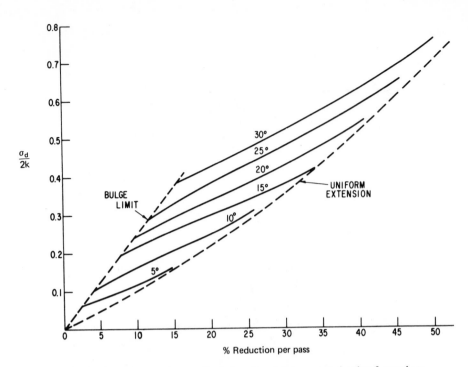

Figure 9-24 Variation of normalized drawing stress versus reduction for various semi-die angles. Note the limit where bulging is predicted. From L. F. Coffin and H. C. Rogers, *Trans. ASM*, 60 (1967), pp. 672–86, as adapted from Hill [2].

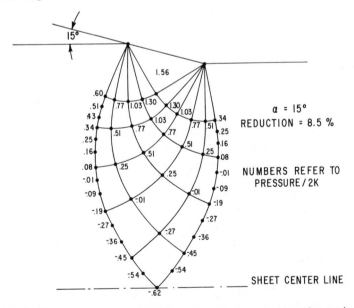

Figure 9-25 Variation in the hydrostatic pressure or mean-normal stress in the deformation zone for strip drawing based upon the field shown. Note that negative values are tensile here. From L. F. Coffin and H. C. Rogers, *ibid.*

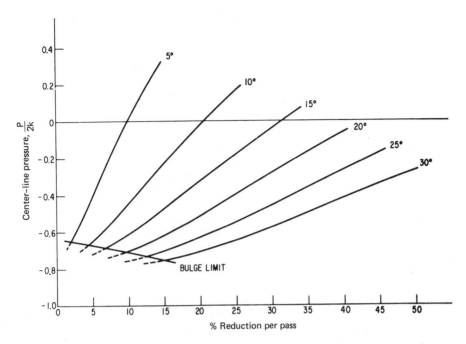

Figure 9-26 Variation in pressure at the centerline for various combinations of r and α during strip drawing. Negative values imply hydrostatic tension. From L. F. Coffin and H. C. Rogers, *ibid.*

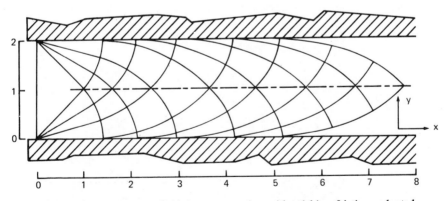

Figure 9-27 A slip-line field for compression with sticking friction, adapted from reference [1].

It should be noted that slip-line fields can be applied to problems with interface conditions intermediate between frictionless and sticking-friction conditions as long as the shear stress, τ, at the interface between tool and workpiece is a constant fraction, m, of the yield stress in shear, k, i.e.,

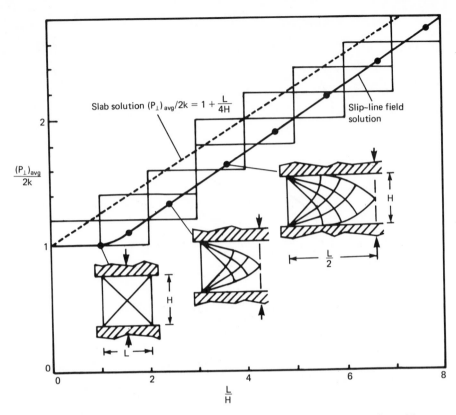

Figure 9-28 Average indentation pressure versus L/H for compression with sticking friction, using the appropriate portion of the slip-line field in Fig. 9-27 and the slab-force analysis given by Eq. (7-25).

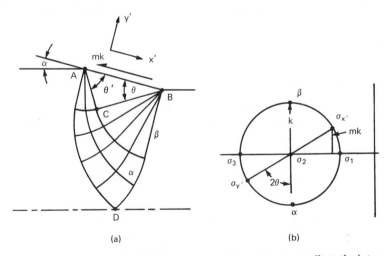

Figure 9-29 (a) Slip-line field where a constant shear stress prevails at the interface and (b) Mohr's stress circle for such conditions.

$$\tau = mk$$

In this case the α- and β-lines meet the tool-metal interface at angles of $\theta = \frac{1}{2}$ arc cos (m) and $\theta' = 90 - \theta$ as indicated in Fig. 9-29a. A general Mohr's circle plot is shown in Fig. 9-29b.

Note that if the conditions are such that in comparison with Eq. (9-22), $r > 2 \sin \alpha/(1 + 2 \sin \alpha)$, a slip-line field different from either Figs. 9-12 or 9-21 must be constructed. This is not done here but is discussed in detail by Johnson and Mellor [1].

9-10 LOAD VARIATION AND PIPE FORMATION IN EXTRUSION

Figure 9-30 shows the general shape of the typical load-displacement diagrams observed for direct and indirect extrusion. Figure 9-12 and the figure for Prob. 9-6 are schematic illustrations of these two processes. There are at least three distinct regions associated with direct extrusion that can be rationalized on a physical basis. The initial and rapid rise in load from O to A is due to the initial

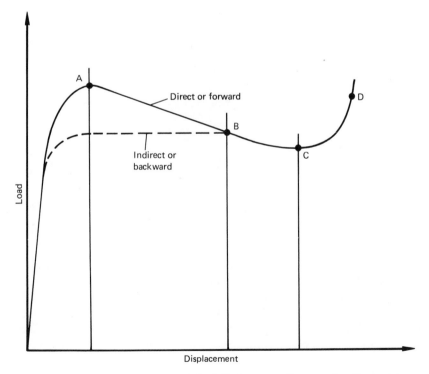

Figure 9-30 Schematic of typical load-displacement diagrams for direct and indirect extrusion.

compression of the billet to fill the container, plus some extrusion of relatively undeformed material. When the load reaches point A, metal begins to exit from the die under steady-state conditions. The drop in load from A to B results from a steadily decreasing frictional effect at the billet-container interface. As more metal is extruded, the contact area at that interface steadily decreases, thereby lowering the total frictional force. Note that with indirect extrusion there is no relative motion at the billet-container interface and, if frictional effects at the billet-piston interface are constant, the steady-state load is constant as shown. At point B, a non-steady state situation begins with an accompanying load drop. Somewhere between B and C (it has been argued that C is the crucial point) a small cavity, called a pipe, forms at the centerline on the back face of the billet. Figure 9-31 is a picture of a pipe.

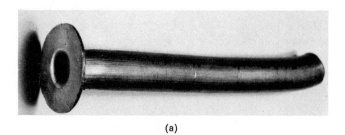

(a)

(b)

Figure 9-31 (a) A billet of lead made by direct extrusion showing a pipe at the left end and (b) a section showing the depth of pipe penetration. Courtesy of W. H. Durrant.

Pipe formation occurs when the radial compressive stresses become large enough to induce buckling of the disc-like material still contained within the extrusion chamber. As displacement continues, the pipe becomes larger. Eventually, the remainder of the billet, now in an annular-like form, is subjected to direct compression, and a rapid load increase, illustrated by point D, occurs. The operation must cease before this load becomes too large or equipment damage will result.

Avitzur [8] has addressed pipe formation and load or pressure variation using an upper-bound approach. Problem 9-13 poses an alternative approach where the steady-state phase is analyzed using a slip-line field; the beginning of the non-steady state condition is treated with a simple upper bound. Problem 9-19, employing a simple upper bound, considers the effect of pipe formation. If the calculations of $P/2k$ are plotted against displacement for these three distinct solutions, and the *lowest* value is used at any particular displacement, the results are similar to Fig. 9-30. Note that since those problems involved frictionless conditions at both the billet-piston and billet-container interfaces, no load drop, such as from A to B on Fig. 9-30, is predicted. It should also be realized that once a pipe forms, the term extrusion pressure loses meaning since the load is no longer uniform across the piston.

REFERENCES

[1] W. Johnson and P. B. Mellor, *Engineering Plasticity*. New York: Van Nostrand Reinhold, 1973, pp. 381–89, 392, 402–6.

[2] R. Hill, *Plasticity*. Oxford: Clarendon Press, 1950, pp. 35–36, 128–160, 166–168, 254, 257.

[3] E. G. Thomsen, C. T. Yang, and S. Kobayashi, *Mechanics of Plastic Deformation in Metal Processing*. New York: Macmillan, 1965, pp. 180–88.

[4] W. Johnson, R. Sowerby, and J. B. Haddow, *Plane-Strain Slip-Line Fields: Theory and Bibliography*. New York: American Elsevier, 1970.

[5] H. Geiringer, *Proc. 3rd. Int. Congr. Appl. Mech. Stockholm*, 2, (1930), pp. 185–190.

[6] A. P. Green, *Phil. Mag.*, 42 (1951), p. 900.

[7] L. F. Coffin and H. C. Rogers, *Trans. ASM*, 60 (1967), pp. 673–74.

[8] B. Avitzur, *Metal Forming: Processes and Analysis*. New York: McGraw-Hill, 1968, pp. 250–74.

APPENDIX

The coordinates y and x of a 5°-net for the slip-line field determined by two centered fans are given in Table 9-1. These were calculated from tables† where the net is described by a different coordinate system. Here the fans have a radius of $\sqrt{2}$, the nodes are separated by a distance of 2, and the origin is mid-way between the nodes, as shown in Fig. 9A-1, for Table 9-1.

†E. G. Thomsen, C. Y. Yang, and Shiro Kobayashi, *Mechanics of Plastic Deformation in Metal Processing* (New York: Macmillan, 1965), pp. 476–79.

TABLE 9-1 Coordinates of a 5° Net for the Slip-Line Field Determined by Two Centered Fans

$\Delta\phi_\alpha$	n	$m=n$	$m=n+1$	$m=n+2$	$m=n+3$	$m=n+4$	$m=n+5$	$m=n+6$	$m=n+7$	$m=n+8$
0°	0	$y=1.000$	1.0833	1.1584	1.2247	1.2817	1.3288	1.3660	1.3926	1.4087
		$x=0.0$	.0910	.1888	.2929	.4023	.5163	.6340	.7544	.8767
5°	1	1.1826	1.2741	1.3572	1.4312	1.4951	1.5484	1.5907	1.6214	1.6399
		0.0	.1000	.2083	.3243	.4472	.5762	.7101	.8484	.9897
10°	2	1.3831	1.4845	1.5770	1.6597	1.7320	1.7925	1.8407	1.8760	1.8975
		0.0	.1106	.2312	.3613	.4999	.6463	.7995	.9583	1.1218
15°	3	1.6050	1.7177	1.8214	1.9146	1.9963	2.0653	2.1206	2.1611	2.1861
		0.0	.1232	.2582	.4046	.5617	.7285	.9038	1.0868	1.2760
20°	4	1.8519	1.9781	2.0946	2.2001	2.2929	2.3718	2.4351	2.4820	2.5108
		0.0	.1377	.2898	.4554	.6339	.8243	1.0257	1.2307	1.4562
25°	5	2.1283	2.2701	2.4018	2.5215	2.6272	2.7176	2.7905	2.8446	2.8781
		0.0	.1550	.3267	.5146	.7181	.9364	1.1680	1.4118	1.6665
30°	6	2.4390	2.5991	2.7484	2.8846	3.0056	3.1093	3.1934	3.2560	3.2948
		0.0	.1749	.3698	.5839	.8166	1.0610	1.3340	1.6162	1.9119
35°	7	2.7897	2.9713	3.1413	3.2968	3.4356	3.5549	3.6519	3.7245	3.7696
		0.0	.1984	.4200	.6647	.9314	1.2196	1.5278	1.8547	2.1985
40°	8	3.1874	3.3940	3.5879	3.7662	3.9257	4.0632	4.1755	4.2595	4.3121
		0.0	.2257	.4787	.7589	1.0655	1.3977	1.7540	2.1332	2.5331
45°	9	3.6394	3.8755	4.0976	4.3023	4.4859	4.6447	4.7747	4.8723	4.9335
		0.0	.2575	.5472	.8688	1.2219	1.6054	2.0182	2.4586	2.9243
50°	10	4.1561	4.4259	4.6808	4.9162	5.1281	5.3117	5.4626	5.5760	5.6472
		0.0	.2947	.6272	.9973	1.4046	1.8482	2.3269	2.8389	3.3828
55°	11	4.7470	5.0565	5.3496	5.6211	5.8659	6.0786	6.2537	6.3856	
		0.0	.3380	.7205	1.1472	1.6179	2.1318	2.6873	3.2831	
60°	12	5.4248	5.7807	6.1185	6.4321	6.7154	6.9622	7.1657		
		0.0	.3886	.8296	1.3223	1.8670	2.4631	3.1091		
65°	13	6.2043	6.6144	7.0043	7.3671	7.6955	7.9820			
		0.0	.4982	.9573	1.5269	2.1584	2.8505			
70°	14	7.1023	7.5758	8.0267	8.4470	8.8281				
		0.0	.5172	1.1055	1.7658	2.4986				
75°	15	8.1290	8.6864	9.2085	9.6961					
		0.0	.5981	1.2794	2.0455					
80°	16	9.3375	9.9715	10.5771						
		0.0	.6925	1.4827						
85°	17	10.7255	11.4605							
		0.0	.8031							
90°	18	12.3341								
		0.0								

n	$n+9$	$n+10$	$n+11$	$n+12$	$n+13$	$n+14$	$n+15$	$n+16$	$n+17$	$n+18$
0	1.4141	1.4087	1.3926	1.3659	1.3288	1.2816	1.2246	1.1584	1.0833	1.0000
	1.0000	1.1233	1.2456	1.3660	1.4837	1.5977	1.7071	1.8112	1.9090	2.0000
1	1.6463	1.6399	1.6068	1.5892	1.5449	1.4879	1.4189	1.3379	1.2455	
	1.1334	1.2718	1.4222	1.5653	1.7061	1.8434	1.9765	2.1036	2.2240	
2	1.9048	1.8975	1.8751	1.8375	1.7846	1.7163	1.6330	1.5347		
	1.2891	1.4582	1.6282	1.7979	1.9658	2.1305	2.2909	2.4452		
3	2.1946	2.1860	2.1595	2.1151	2.0522	1.9707	1.8707			
	1.4708	1.6686	1.8688	2.0694	2.2690	2.4658	2.6583			
4	2.5207	2.5107	2.4797	2.4272	2.3527	2.2555				
	1.6828	1.9141	2.1497	2.3865	2.6233	2.8574				
5	2.8895	2.8778	2.8414	2.7795	2.6913					
	1.9304	2.2014	2.4777	2.7580	3.0372					
6	3.3083	3.2944	3.2519	3.1789						
	2.2196	2.5366	2.8610	3.1901						
7	3.7853	3.7692	3.7191							
	2.5573	2.9281	3.3089							
8	4.3303	4.3114								
	2.9518	3.3859								
9	4.9548									
	3.4129									

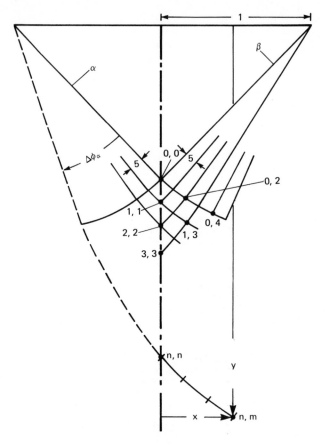

Figure 9A-1

PROBLEMS

9-1. Using the slip-line field in Fig. 9-9 for frictionless, plane-strain indentation, $P/2k$ was found to be 2.57 as in Eq. (9-19). The sketch shows an alternate field proposed by Hill [2].

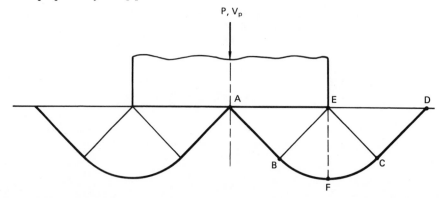

(a) Find $P/2k$ for this field if frictionless conditions prevail along AE.

(b) Construct a hodograph to scale.

(c) What percent of the energy input is dissipated along the lines of intense shear?

9-2. The sketch shows a proposed slip-line field connected with a frictionless punch used to indent a metal in plane strain. Construct an appropriate hodograph to scale and find $P/2k$.

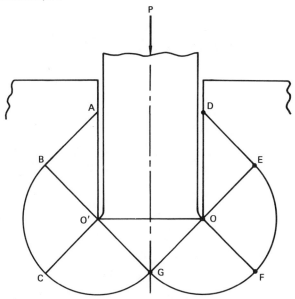

9-3. Plane-strain compression of a hexagonal rod is shown in the sketch, including a possible slip-line field. The metal has a tensile yield strength of 25,000 psi. If

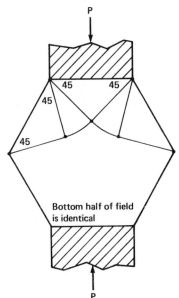

the von Mises yield criterion applies, find the pressure P, in psi, to start flow of the metal.

9-4. Asymmetric, frictionless, plane-strain extrusion is indicated in the sketch and a proposed field is shown.

    (a) Construct a hodograph to scale to determine the magnitude and direction of the exit velocity.

    (b) Discuss the physical appropriateness of this field.

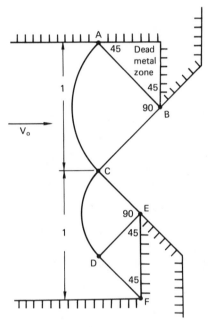

9-5. A slip-line field for wedge indentation of a semi-infinite slab consists of two centered fans and four constant pressure zones as shown. The angle ψ is related

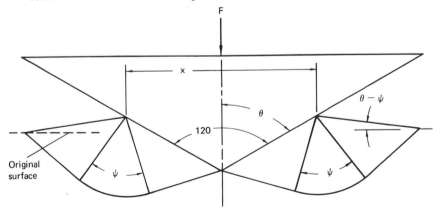

to θ by the relation $\cos(2\theta - \psi) = \cos\psi/(1 + \sin\psi)$, in order that the volume of the side mounds equals the volume displaced by the indentor; note that for this case $2\theta = 120°$. Determine the ratio F/x in terms of $2k$.

9-6. A 2:1 reduction is to be performed by indirect or backward extrusion as shown; frictionless, plane-strain conditions prevail. The slip-line field shown consists of two equal centered fans.
(a) Determine $P_e/2k$.
(b) Construct a hodograph to scale for the lower half of the field.
(c) What is the velocity discontinuity along AD? along CD?
(d) What percentage of the total energy input is dissipated by the gradual deformation of metal moving through the centered fan?

ADE and A'D'E' are dead metal zones

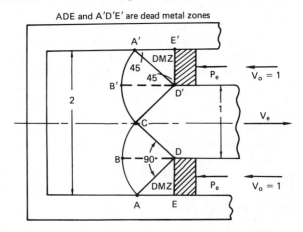

9-7. A 3:1 reduction is to be accomplished by frictionless, plane-strain extrusion and a proposed field is shown. Find $P_e/2k$, then determine the deformation efficiency, η, for this process.

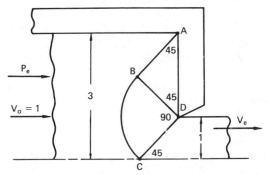

9-8. The indenting of an infinite slab containing a step is shown below along with a suggested slip-line field.
(a) With proper calculations, show that this full field cannot be correct.
(b) Suggest an alteration that would produce a correct field.

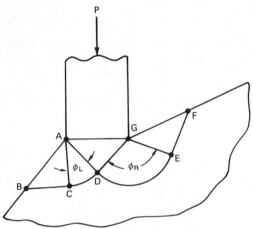

9-9. The conditions here are identical to those in Prob. 9-7, where the slip-line field is *ABECD*. Added here is an upper-bound field consisting of triangles *ABD* and *BCD*. Points *E*, *F*, and *D* lie on a line that is 45° from *BD* or *CD*.

(a) Draw the hodograph for each field.

(b) Determine the absolute velocity at a point just to the right of *E* in the slip-line field.

(c) Determine the absolute velocity of *F* in the upper-bound field.

(d) Compare the energy dissipated along *BC* of the upper bound with that along arc *BEC* of the slip-line field.

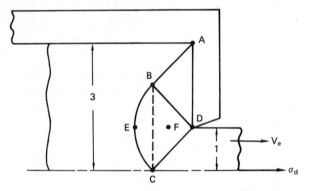

9-10. Frictionless, plane-strain extrusion produces the reduction shown by the use of

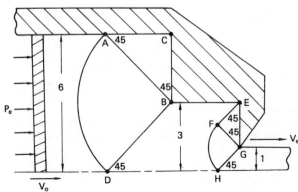

a two-stage die. The two regions utilize the slip-line fields shown, where ACB is
a dead-metal zone.

(a) Find $P_e/2k$.

(b) What is the velocity of a particle in region EFG?

9-11. Back extrusion, under frictionless, plane-strain conditions is shown. For the
field,

(a) Find $P_e/2k$.

(b) Construct a hodograph to scale.

(c) What is the exit velocity, V_e?

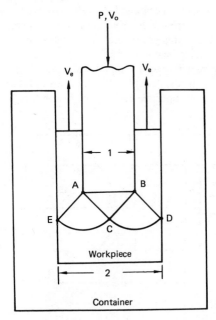

9-12. Consider punching a long thin slot of width, w, in a metal sheet of thickness t
as shown in the sketch. For punching, shear must occur along lines AB and CD.

(a) Using an upper-bound solution, find $P/2k$ as a function of t.

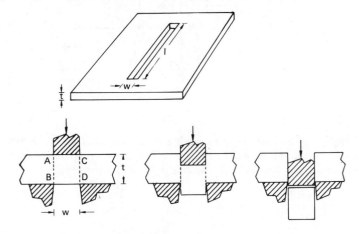

(b) If the ratio of t/w is too great, attempting to punch will result in a two-dimensional hardness test. What is the maximum value of t/w for punching?

(c) Consider punching of a circular hole of diameter, d. Assume that for a hardness indentation, $P/2k = 3$. What is the largest ratio of hole diameter-to-sheet thickness that can be punched?

9-13. An analysis that combines a slip-line field and an upper bound is to be used for direct extrusion under frictionless and plane-strain conditions. Sketch (a) shows the slip-line field which is appropriate from the onset of steady-state flow until the distance t becomes small enough that the piston itself would alter the field. Once that happens, the upper bound in sketch (b) can be used, and as the piston continues to move, P varies with t. This is a non-steady state condition. (Although not to be considered in this problem, it is of interest to note that when t becomes small enough, a cavity begins to form at the centerline of the metal not yet extruded; this is referred to as a pipe.) Calculate $P_e/2k$ as a function of t for $\frac{1}{4} \leq t \leq 5$ for both fields, and plot $P_e/2k$ versus t. Indicate the range of t for which each field is appropriate. Assume the piston face is frictionless.

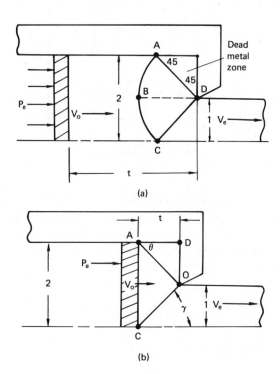

(a)

(b)

9-14. A deeply notched tensile specimen is shown, the specimen being much longer than its width and very deep in a direction perpendicular to the plane shown. For the slip-line field drawn, calculate $\sigma/2k$, where $\sigma = F/x$.

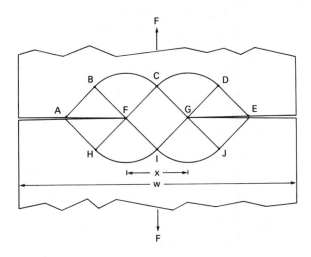

9-15. A plane-strain tension test is to be conducted using the notched specimen shown in sketch (a), where $w \gg t_o$. The metal is ideally plastic, the notch angle is 90°, and the radius at the root of the the the notch is negligibly small. Using the slip-line field shown in sketch (b), calculate $\sigma_x/2k$ to cause yielding, where $\sigma_x = F_x/wt_n$.

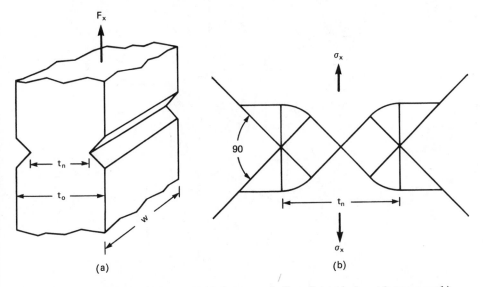

9-16. Regarding Prob. 9-15, if the notches are too shallow (i.e., t_o/t_n is not large enough) an alternate mode of deformation will occur as shown in the sketch, with shearing taking place on planes extending from the base of one notch to the opposite face. Calculate $\sigma_x/2k$ as a function of t_o/t_n if this mode occurs. What minimum ratio of t_o/t_n is necessary to prevent this mode?

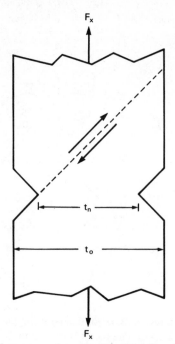

9-17. The sketch shows the appropriate slip-line field for either extrusion or drawing

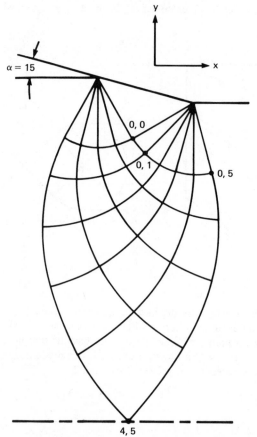

under plane strain, where $r = 0.0760$, $\alpha = 15°$, and frictionless tools are used. A solution, with extrusion, gives $P_e/2k = 0.178$.

(a) Find the level of the hydrostatic stress, $\sigma_2 = (\sigma_1 + \sigma_2 + \sigma_3)/3$ at point 4, 5 during extrusion.

(b) Repeat part (a) for a drawing operation.

(c) How might the structure and properties of the final product differ depending upon whether drawing or extrusion is used?

9-18. A long rod, whose cross section is a regular octagon, is compressed between flat, frictionless platens; all metal flow is in the plane of the sketch. Two slip-line fields are proposed, one indicating "penetrating" deformation, while the other is a "modified hardness" indentation. Which field is more appropriate? Justify your answer with appropriate calculations and reasoning.

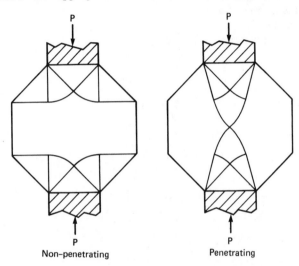

Non-penetrating   Penetrating

9-19. In Prob. 9-13, the direct extrusion pressure, under frictionless conditions, could be analyzed using a slip-line field and an upper bound. Depending upon the position of the piston, one of these solutions would predict a lower value for $P_e/2k$. As discussed in Sec. 9-10, pipe formation occurs as the piston approaches the end of the container. In the sketch, point C is *on* the piston face (i.e., a pipe has started to form); note that θ and the position of C vary with t. Take P_e as

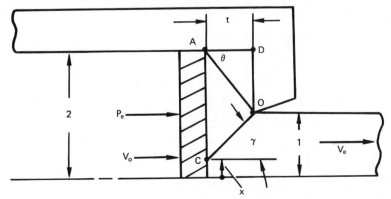

the extrusion *force* divided by the area of the piston. Determine $P_e/2k$ as a function of t for this deformation mode and compare with the results in Prob. 9-13.

9-20. Consider an extrusion with a frictionless die and $\alpha = 30°$ such that the point 2, 4 on Fig. 9-19 is at the centerline.

   (a) What is the reduction?

   (b) Calculate the value of $P_e/2k$.

   (c) What is the deformation efficiency?

   (d) Find the mean hydrostatic stress σ_2 at the centerline.

9-21. Consider the slip-line field for drawing with a constant shear stress, $\tau = mk$, along the die wall. Using Fig. 9-29, let $\alpha = 15°$ and $\theta = 30°$. Point D corresponds to point 2, 4 in Fig. 9-19.

   (a) What is the value of m?

   (b) What is the reduction?

   (c) Calculate $\sigma_d/2k$.

   (d) What is the deformation efficiency?

Deformation-zone geometry

10-1 THE Δ-PARAMETER

As shown previously, the shape of the deformation zone exerts a strong influence upon redundant work, frictional work, and total forming forces in drawing, extrusion, and rolling. It also has important effects on the properties and structure after forming; these include homogeneity of hardness, internal porosity, tendency to open cracks during processing, and residual stresses. These effects, which are common to drawing, extrusion, and rolling, will be reviewed in this chapter. Deformation zone geometry for all of these processes can be characterized by a single parameter, Δ, defined as the ratio of the mean thickness or diameter, h, of the work metal to the contact length between tool and work metal,

$$\Delta = \frac{h}{L} \qquad (10\text{-}1)$$

This parameter is useful in correlating and summarizing, albeit not perfectly, many effects of the deformation zone.

For drawing or extrusion, the contact length, $L = (h_o - h_1)/(2 \sin \alpha)$, and the mean thickness or diameter, $h = (h_o + h_1)/2$, so

$$\Delta = \frac{(h_o + h_1)}{(h_o - h_1)} \sin \alpha \qquad (10\text{-}2)$$

In plane-strain drawing or extrusion, the reduction is given by $r = (h_o - h_1)/h_o$, so

$$\Delta = \frac{(2 - r)}{r} \sin \alpha \qquad (10\text{-}3)$$

For axisymmetric drawing, Eq. (10-2) holds where h is the diameter; thus,

$r = (d_o^2 - d_f^2)/d_o^2$, so

$$\Delta = \left(\frac{1 + \sqrt{1 - r}}{1 - \sqrt{1 - r}}\right) \sin \alpha = \frac{\sin \alpha}{r}(1 + \sqrt{1 - r})^2 \qquad (10\text{-}4)\dagger$$

For flat rolling, the evaluation of Δ is much simplified if we use the chordal length of the contact arc, $L = \sqrt{R\Delta h} = \sqrt{rRh_o}$ instead of the actual arc of contact, and for most rolling conditions this simplification causes little error. Since the mean thickness is $h = (h_o + h_1)/2 = h_o(1 + h_1/h_o)/2 = h_o(2 - r)/2$,

$$\Delta = \frac{(2 - r)}{2} \sqrt{\frac{h_o}{rR}} \qquad (10\text{-}5)$$

Other authors‡ use slightly different definitions of Δ§.

In all of these formulae, Δ increases with decreasing reduction and increasing die angle (or in rolling with the ratio of strip thickness to roll radius).

10-2 FRICTION

The frictional contribution to the total work, w_f/w_a, increases with decreasing Δ, because as Δ decreases, the contact area between the workpiece and tool increases; since the normal force acting between the workpiece and tool is proportional to the contact area, the frictional forces increase.

In both plane-strain and axisymmetric drawing, the ratio of die-workpiece contact area to mean cross-sectional area** of the workpiece is

$$\text{Area ratio} = \frac{2r}{(2 - r)\sin \alpha} \qquad (10\text{-}6)$$

which for plane strain is $2/\Delta$ and for axisymmetry is equal to about $4/\Delta$.

The upper-bound slab analysis for a constant shear-stress interface can be used to predict the frictional contribution to the drawing load. Taking $\sigma_o = 2k$ in Eqs. (8-39) and (8-43), $w_h = 2k\epsilon_h$ and $w_f = 2k\epsilon_h m/\sin 2\alpha$, so

$$\frac{w_f}{w_h} = \frac{m}{\sin 2\alpha} \qquad (10\text{-}7)$$

†It should be noted that for the same α and r, Eq. (10-4) gives a Δ-value for axisymmetry approximately twice that for plane strain, Eq. (10-3). It might have been better to define the Δ-parameter in terms of the ratio of the mean cross-sectional area of the workpiece to the tool-workpiece contact area. Such a definition would give the same Δ-values for axisymmetric and plane-strain operations having the same α and r. However, we feel that such a deviation from the conventional definitions of Δ would create confusion.

‡Backofen (*Deformation Processing*, Reading, Mass.: Addison-Wesley, 1972) defines the mean height, h, as the length of an arc, normal to the tools drawn through the center of the deformation zone. His expressions for plane-strain and axisymmetric drawing and extrusion are identical to Eqs. (10-2), (10-3), and (10-4) except that $\sin \alpha$ is replaced by α (in radians). Unless α is large, $\alpha \simeq \sin \alpha$, so the definitions are similar.

§R. M. Caddell, and A. G. Atkins, *J. Eng. Ind., Trans. ASME*, Series B, 90, no. 2 (May 1968), pp. 411–19.

**The mean cross-sectional area is $(A_o + A_f)/2$.

for both axisymmetry and plane strain. For small values of α, $\sin 2\alpha \approx 2 \sin \alpha$, so at constant reduction $w_f/w_h \propto 1/\Delta$.

With a constant coefficient of friction, Eq. (7-14) for axisymmetry and Eq. (7-10) for plane strain both give

$$\frac{\sigma_d}{\sigma_o} = \frac{1+B}{B}[1 - \exp(-B\epsilon_h)] \tag{10-8}$$

where $B = \mu \cot \alpha$. Substituting the series expansion, $\exp(-x) = 1 - x + x^2/2 \ldots$, $\sigma_d/\sigma_o = \epsilon(1 + B - B^2\epsilon/2 + \ldots)$. Now substituting $w_h = \sigma_o\epsilon$ and $w_f = \sigma_d - w_h$,

$$\frac{w_f}{w_h} = B\left(1 - \frac{\epsilon}{2} + \cdots\right) \simeq \mu \cot \alpha \left(1 - \frac{\epsilon}{2}\right) \tag{10-9}$$

For reasonably low die angles, $\cot \alpha \simeq 1/\sin \alpha$, so, referring to Eqs. (10-3) and (10-4), again $w_f/w_h \propto 1/\Delta$ for constant r.

10-3 REDUNDANT DEFORMATION

It is convenient to describe the redundant strain, ϵ_r, by a factor $\phi = (\epsilon_r + \epsilon_h)/\epsilon_h$, where ϵ_h is the homogeneous strain. In the absence of work hardening, $(w_r + w_h)/w_h = (\epsilon_r + \epsilon_h)/\epsilon_h = \phi$, so in drawing, the drawing stress can be expressed as $\sigma_d = \phi\sigma_o\epsilon_h + w_f$, where w_f is the frictional work. Therefore

$$\phi = \frac{\sigma_d - w_f}{\sigma_o\epsilon_h} \tag{10-10}$$

Equations (8-39) and (8-43), from the upper-bound slab analysis, can be used to predict ϕ for plane-strain and axisymmetric drawing. Taking $\sigma_o = 2k$ in those equations, and comparing with Eq. (10-10),

$$\phi = 1 + \frac{\frac{1}{2}\tan \alpha}{\epsilon_h} \tag{10-11a}$$

for plane strain and

$$\phi = 1 + \frac{\frac{2}{3}\tan \alpha}{\epsilon_h} \tag{10-11b}$$

for axisymmetry. Numerical evaluation of Eqs. (10-11a) and (10-11b) for various combinations of α and $\epsilon_h = \ln(1/1 - r)$, and with $\alpha < 30°$ and $r < 0.5$ results in $\phi \simeq 1 + \Delta/4$ for plane strain and $\phi \simeq 1 + \Delta/6$ for axisymmetry.

The slip-line field solution for plane-strain indentation (Fig. 9-20) can be approximated by a straight line with the equation

$$\phi = 1 + 0.23(\Delta - 1) \quad \text{for} \quad 1 \leq \Delta \leq 8.8 \tag{10-12}$$

At present, there is no corresponding solution for axisymmetry.

There have been several experimental studies relating the redundant strain, ϕ, to process variables. The most direct method of determining ϕ is to compare the true stress-strain curves before and after deformation. Figure 10-1, from Caddell and Atkins[†], shows the tensile stress-strain curves of 303 stainless

†R. M. Caddell and A. G. Atkins, *J. Eng. Ind., Trans. ASME*, Series B, 90 (1968), p. 411–19.

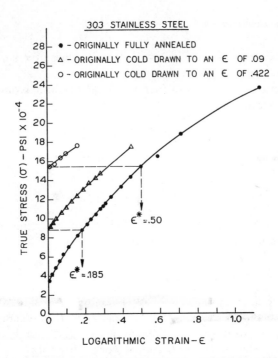

Figure 10-1 Stress-strain curves for 303 stainless steel in the annealed state and after drawing. From R. M. Caddell and A. G. Atkins, *J. Eng. Ind., Trans. ASME,* Series B, 90 (1968) pp. 411–19.

steel in the annealed condition, and after rod drawing to ideal strains of $\epsilon_h = 0.090$ and 0.422. Comparing the curves shows that the yield strength of the bar drawn to $\epsilon_h = 0.090$ is equivalent to the flow stress after a tensile strain of 0.185. Therefore, the actual strain induced by drawing must have been $\epsilon_a = 0.185$ and $\phi = 0.185/0.090 = 2.05$. Similarly, for the heavier reduction $\phi = 0.500/0.422 = 1.18$. With various die angles and reductions they explored a wide range of Δ-values. Figure 10-2 shows how the experimental values of ϕ varied with Δ. For three different materials the trends can be described by

$$\phi = C_1 + C_2\Delta \tag{10-13}$$

where C_1 and C_2 are constants.

A similar method is based upon hardness measurements. First it is determined how hardness varies with strain during homogeneous deformation (tension test or rolling with a geometry chosen to insure homogeneity). Then the hardnesses after various cold-forming operations are compared to those after homogeneous deformation to determine ϵ and ϕ. Backofen[†] reported such a study[‡] for both axisymmetric and plane-strain drawing. The results in Fig. 10-3 are similar except the values of C_1 and C_2 vary. On physical grounds

[†]W. A. Backofen, *ibid.*

[‡]J. J. Burke, ScD Thesis, Dept. of Metallurgy and Materials Science, MIT (1968).

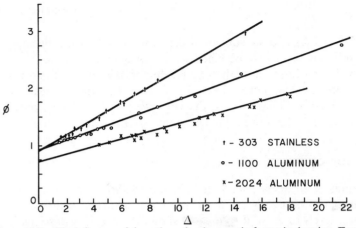

Figure 10-2 The influence of Δ on the redundant strain factor in drawing. From R. M. Caddell and A. G. Atkins, *ibid.*

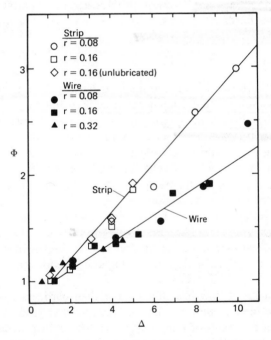

Figure 10-3 The influence of Δ on the redundant strain factor for axisymmetric and strip drawing in copper. Adapted from W. A. Backofen, *Deformation Processing* (Reading, Mass.: Addison-Wesley, 1972), p. 140, Fig. 7-5a; and J. J. Burke, ScD Thesis, MIT (1968).

ϕ can never be less than unity, so Eq. (10-13) must have a lower cutoff. For that reason Backofen suggested that

$$\phi = 1 + C(\Delta - 1) \quad \text{for} \quad \Delta \geq 1 \quad \text{and} \quad \phi = 1 \quad \text{for} \quad \Delta \leq 1 \qquad (10\text{-}14)$$

where $C = 0.21$ for plane-strain and 0.12 for axisymmetric deformation. The value of $C = 0.21$ is quite similar to the value of 0.23 from the slip-line field, though other workers have obtained somewhat different values.

It is interesting to note that the results for axisymmetric and plane-strain deformation would be much closer if ϕ had been plotted against the ratio of the mean cross-sectional area of the workpiece to the die-workpiece contact area. This ratio is $\Delta/4$ for axisymmetry and $\Delta/2$ for plane strain.

10-4 INHOMOGENEITY

The redundant strain is not evenly distributed throughout the cross section. Figure 10-4 shows the distortion of grid lines calculated from slip-line theory for several levels of Δ, while Fig. 10-5 shows the actual distortions in gridded billets after axisymmetric extrusion. Thus, both theory and experiment show that redundant strain increases with Δ and is greatest near the surface. This too is indicated in Fig. 10-6 by hardness profiles across rolled strip. Backofen[†] characterized the hardness gradients in wire and strip drawing experiments by an inhomogeneity factor defined as

$$\text{I.F.} = \frac{H_s - H_c}{H_c} \tag{10-15}$$

where H_s and H_c are the Vickers hardness at the surface and center. His plots of I.F. versus Δ indicate that I.F. increases with Δ but the variations of I.F. are not completely described by Δ alone: at constant Δ-levels, I.F. increases with α and decreases with r. Furthermore, at the same Δ and r, the I.F. for strip drawing (plane strain) is roughly twice as great as for wire drawing (axisymmetry). However, when his data are replotted as I.F. versus α as in Fig. 10-7, plane-strain and axisymmetric points are brought into close accord and are much less dependent on reduction.

Hardness variations at larger reductions and die angles were studied in hydrostatic extrusion experiments by Miura et al.[‡] Figure 10-8 shows how the difference between surface and centerline hardness increases with both increasing die angle and decreasing reduction. At the higher reductions the results correlate roughly with Δ.

Inhomogeneity of flow has effects other than hardness variations. There is considerable evidence[§] showing that surface-to-center gradients of crystallographic texture are a result of inhomogeneous flow. When plates or sheets are rolled under low-Δ conditions, such textural gradients disappear. Furthermore, since the grain size after recrystallization depends upon prior strain, higher surface strain can result in a finer recrystallized grain size on the surface than in the center if the workpiece is annealed after cold working. Finally, if the deformation is rapid enough to be adiabatic, gradients in strain will cause more

†W. A. Backofen, *ibid*.

‡S. Miura, Y. Saeki, and T. Matushita, *Metals and Materials*, 7 (1973), pp. 441–47.

§P. S. Mathur and W. A. Backofen, *Met. Trans.*, 4 (1973), pp. 643–51.

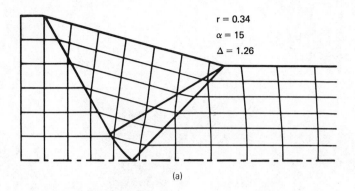

(a)

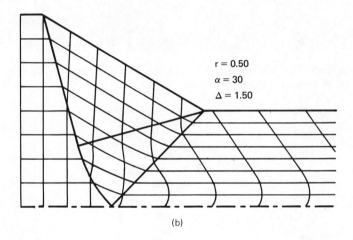

(b)

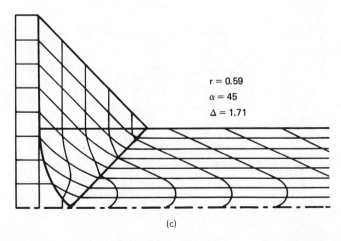

(c)

Figure 10-4 Grid distortions predicted from slip-line field theory for strip drawing. Note that the distortion increases with Δ.

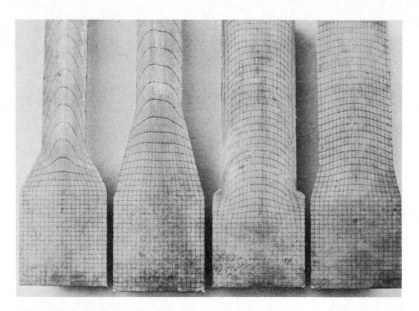

Figure 10-5 Grid distortion of cold-extruded billets. From D. J. Blickwede, *Metals Progress*, 97 (May 1970), pp. 76–80.

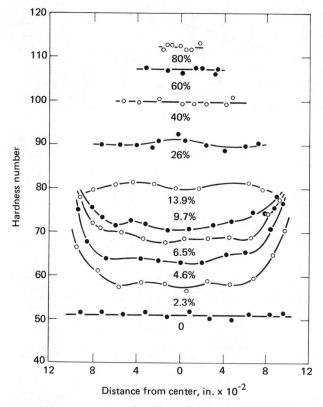

Figure 10-6 Hardness gradients in copper after rolling in one pass to indicated reduction. Initial strip thickness was 0.2 in. and the roll diameter was 10 in. From B. B. Hundy and A. R. E. Singer, *J. Inst. Metals*, 83 (1954–5), pp. 401–7.

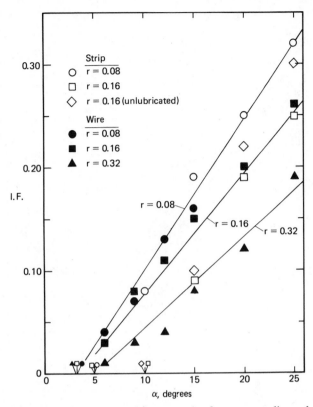

Figure 10-7 Dependence of the inhomogeneity factor upon die angle. Data from W. A. Backofen and J. J. Burke, *ibid*.

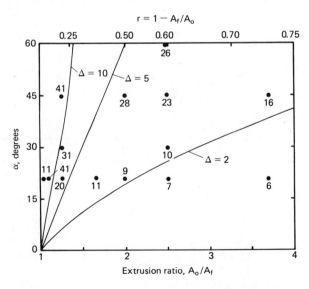

Figure 10-8 Effect of die angle and reduction on inhomogeneity of steel extrusions. The numbers are the difference between surface and centerline hardness. Lines of constant Δ were calculated from Eq. (10-4). Data from Miura et al., *ibid*.

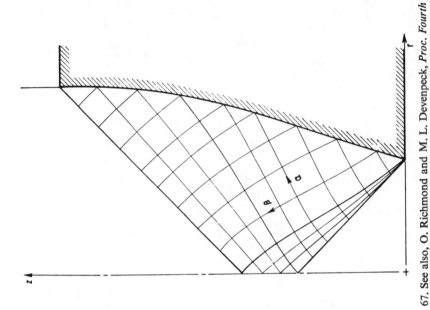

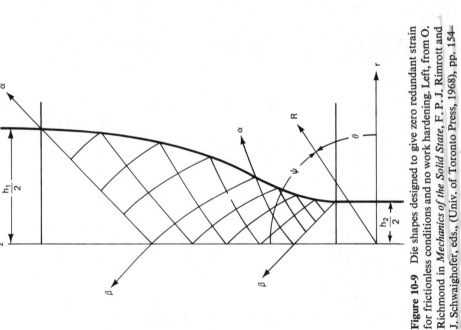

Figure 10-9 Die shapes designed to give zero redundant strain for frictionless conditions and no work hardening. Left, from O. Richmond in *Mechanics of the Solid State*, F. P. J. Rimrott and J. Schwaighofer, eds., (Univ. of Toronto Press, 1968), pp. 154–67. See also, O. Richmond and M. L. Devenpeck, *Proc. Fourth U.S. Nat. Cong. of Appl. Mech.* (1962), p. 1053. Right, from O. Richmond and H. L. Morrison, *J. Mech. Phys. Solids*, 15 (1967), pp. 195–203.

heating and higher temperatures near the surface than in the interior. As will be indicated later, this can affect residual stresses after deformation.

Throughout the discussion above it has been assumed that the dies are conical or flat wedges. It is interesting to note that Devenpeck and Richmond† designed plane-strain dies (using slip-line theory) and axisymmetric dies (by numerical methods) which, in principle, produce no redundant strain and, therefore, no surface-to-center property gradients. Figure 10-9 gives two such profiles and Fig. 10-10 shows that the distortion of grid lines is small. However, it should be noted that these dies are characterized by a low Δ, so large effects would not be expected in such dies.

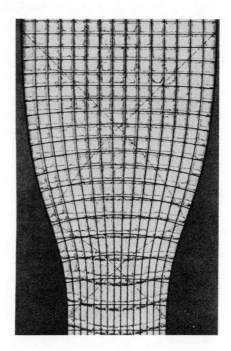

Figure 10-10 Overlay of actual and theoretical grid distortions with die similar to Fig. 10-9 (left). From M.L. Devenpeck and O. Richmond, *J. Eng. Ind.*, *Trans. ASME*, 87 (1965), p. 425.

10-5 INTERNAL DAMAGE

Processing under high-Δ conditions can lead to internal damage in the form of microscopic porosity near the centerline, or in extreme cases, internal cracks. With high-Δ deformation zones there may be a state of hydrostatic *tension* near the centerline. Note that hydrostatic tension is not limited to conditions of drawing since it may exist during rolling or extrusion.

Figure 9-25 shows the slip-line field for a low reduction, high-α, plane-strain drawing (Δ = 5.83) and the predicted levels of hydrostatic tension

†M. Devenpeck and O. Richmond; see Fig. 10-9.

($\sigma_2 = -P$) throughout the field. The high level of hydrostatic tension near the mid-plane is evident. Figure 9-26 shows how the level of hydrostatic tension at the mid-plane increases with increasing die angle, α, and drawing reduction, r. Internal damage increases with increasing levels of hydrostatic tension. Coffin and Rogers[†] showed significant decreases in density after slab drawing with high-Δ fields, particularly in the material near the centerline. They found that the density loss is cumulative from pass to pass, and that greater losses occur with smaller reductions per pass and higher die angles; Fig. 10-11 indicates this tendency. The density loss is due to porosity formed on a microscopic level, especially near inclusions or hard second-phase particles, and varies with material. Clean single-phase materials show a much lower decrease in density than dirty materials with second phases. Figure 10-12 compares tough-pitch copper, with a high content of Cu_2O particles, and OFHC, oxygen-free high-conductivity copper. It has also been found that the loss in density could

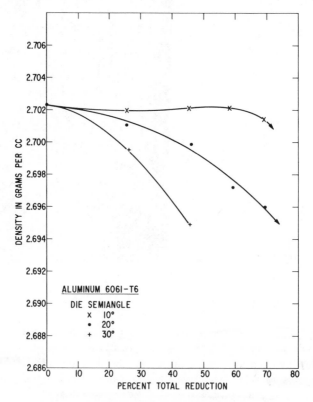

Figure 10-11 Density changes of 6061-T6 aluminum caused by drawing. Note the greater loss of density at high die angles. From H. C. Rogers, *General Electric Co. Report No. 69-C-260* (July 1969).

[†]L. F. Coffin, Jr. and H. C. Rogers, *Trans. ASM*, 60 (1967) pp. 672–86.

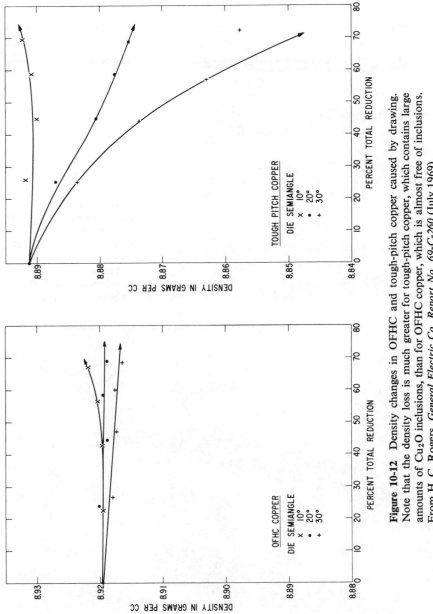

Figure 10-12 Density changes in OFHC and tough-pitch copper caused by drawing. Note that the density loss is much greater for tough-pitch copper, which contains large amounts of Cu_2O inclusions, than for OFHC copper, which is almost free of inclusions. From H. C. Rogers, *General Electric Co. Report No. 69-C-260* (July 1969).

be minimized or prevented by drawing under external hydrostatic pressure. In some cases the strip density actually increased, presumably by decreasing porosity formed in prior processing.

Under severe enough conditions, progressive damage can cause the pores to link up, thereby forming "chevron" cracks along the centerline. The terms "arrowhead cracks," "centerline burst," and "cuppy core" have also been used

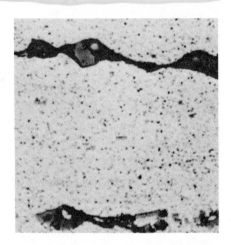

Figure 10-13 Voids formed in 6061-T6 aluminum strip drawn with 28% reduction per pass to a total reduction of 75% through 30° dies (930x, reduced 6% for publication). From H. C. Rogers, R. C. Leech, and L. F. Coffin, Jr., *Final Rept.*, *Contract NOw-65-0097-f*, Bureau of Naval Weapons (1965).

Figure 10-14 Centerline cracks in extruded steel rods. From D. J. Blickwede, *ibid.*

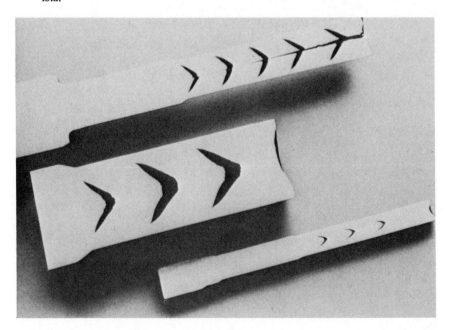

Figure 10-15 Cracks formed in a molybdenum bar rolled under high Δ conditions.

to describe this phenomenon. An early stage of such crack formation is shown in Fig. 10-13, while Figs. 10-14 and 10-15 show cracks formed in extrusion and rod rolling. Even if such cracks don't cause actual separation in processing, they will certainly decrease the conductance and the mechanical properties of the final product. Microscopic porosity can lower the ductility, toughness, and fatigue resistance of the product, even when macroscopic cracks are not present.

Avitzur† has analyzed the problem of central bursts using an upper-bound analysis to find the conditions for which opening a central hole reduces the deformation energy. Figure 10-16 shows the results of his analysis for extrusion and the results for several work-hardened steels. It is interesting to note that the conditions which produce central burst can also be described by Δ > 2. The safe forming region in Fig. 10-16 is larger for materials which work harden rapidly and for materials which are relatively free from inclusions. Yet the general effect of deformation geometry is reasonably described by Δ.

10-6 RESIDUAL STRESSES

The nature of residual stresses after rolling, extrusion, and drawing depends upon the shape of the deformation zone. With Δ-values of one or lower (high reductions, large roll diameter to thickness ratios, or small die angles) the deformation is relatively uniform and residual stresses are minor. In general, as Δ increases above unity the surface is left in residual tension (axial direction stresses) and the center in compression. The magnitude of the stresses increases with Δ as shown in Figs. 10-17, 10-18, and 10-19 for rolling, drawing, and extrusion.

If the extrusion is rapid enough, temperature gradients caused by inhomogeneous deformation may cause an added effect upon subsequent cooling. The hotter surfaces would, if free from the interior, undergo more thermal contraction than the interior. However, since the surface and interior must

†See, for example, B. Avitzur, *Metal Forming: Processes and Analysis* (New York: McGraw-Hill, 1968), pp. 172–176.

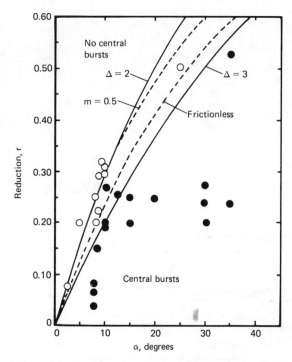

Figure 10-16 The effect of Δ on centerline bursting of three steels during extrusion. The filled circles indicate centerline bursts and open circles indicate no center bursts. The dashed lines are Avitzur's theoretical predictions for zero work hardening with no friction and with an interface shear stress of $mk = 0.5k$. Solid lines are lines of constant Δ calculated from Eq. (10-4). Adapted from Z. Zimmerman and B. Avitzur, *J. Eng. Ind., Trans. ASME*, 92 (1970), p. 135.

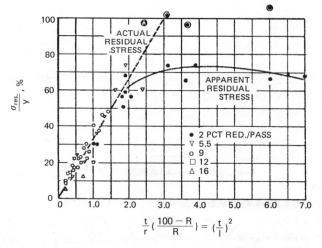

Figure 10-17 Residual stresses at the surface of rolled strip. The residual stresses are normalized by the yield strength and the abscissa is Δ^2. From W. M. Baldwin, *Proc. ASTM*, 49 (1949), pp. 539–83; adapted from R. M. Baker, R. E. Rieksecker, and W. M. Baldwin, *Trans. AIME*, 175 (1948), pp. 337–54.

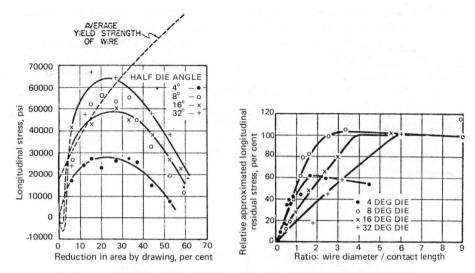

Figure 10-18 Residual stresses at the surface of cold-drawn brass wire (left), and the same data replotted as ratio of residual stress to yield strength versus Δ (right). From W. M. Baldwin, *ibid.* (Original data from W. Linicus and G. Sachs, *Mitt. Dtsch. Materialprüfunsanst*, 16 (1932), pp. 38–67.

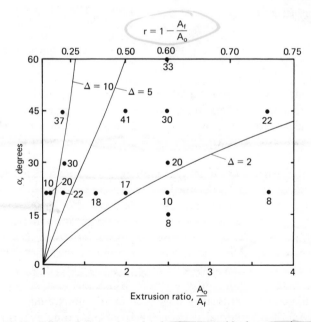

Figure 10-19 Effect of die angle and reduction on residual stresses in extruded steel rods. Numbers are the axial residual stress on the surface in kg/mm², and the constant Δ lines were calculated from Eq. (10-4). Data from S. Miura et al., *ibid.*

undergo equal contraction, the surface would be left under residual tension. Nevertheless, this is not the major cause of residual stresses.

High residual stresses, coupled with damage at the centerline can result in spontaneous splitting or "alligatoring" of the work material as it leaves the deformation zone. An example of alligatoring is shown in Fig. 10-20. This type of failure is most likely in early breakdown rolling of ingots because the section thickness is usually large relative to the roll diameter, while the reduction per pass is low. Furthermore, the structure of the ingot will not have benefited from prior deformation. In cast ingots, porosity and second phases near the center are potential sources of weakness. Hot rolling under favorable conditions will tend to close up such porosity.

Figure 10-20 Molybdenum rod that alligatored during rolling under high-Δ conditions. The "teeth," formed by edge cracks, are not common in alligatored bars.

The normal pattern of residual stresses may be completely reversed if Δ is so large that the deformation zone does not penetrate to the centerline. In that case the surface is left in residual compression as shown in Fig. 10-21. Skin-rolling passes and shot peening are operations where such residual stresses develop.

Despite the general usefulness of slip-line field theory in metal forming, it does not adequately explain observed residual stress patterns. For deformation zones with high Δ-values, slip-line theory predicts that *during deformation*, the level of hydrostatic stress is most tensile at the centerline and most compressive at the surface. The theory allows for no plastic deformation after material leaves the field, so unloading must be entirely elastic and the net axial and transverse forces must go to zero. The normal stress must drop to zero while maintaining longitudinal and traverse strains which are compatible across the section. It is difficult to rationalize how such elastic unloading could cause a complete reversal of the internal stresses. The problem probably lies in the fact that slip-line field theory assumes rigid-perfectly plastic behavior, so there is no strain compatibility problem upon unloading; with real materials, unloading may involve some plastic deformation. An additional factor may be

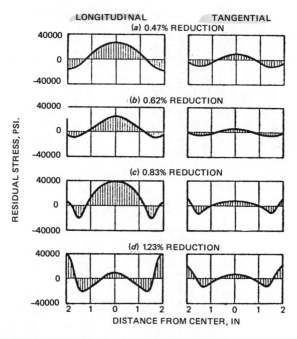

LONGITUDINAL   TANGENTIAL

RESIDUAL STRESS, PSI.

(a) 0.47% REDUCTION

(b) 0.62% REDUCTION

(c) 0.83% REDUCTION

(d) 1.23% REDUCTION

DISTANCE FROM CENTER, IN

Figure 10-21 Residual stresses in lightly drawn steel rods. From W. M. Baldwin, *ibid.*, after H. Bühler and H. Buchholtz, *Archiv. fur des Eisenhüttenwesen*, 7 (1933–4), pp. 427–30.

that while slip-line theory assumes no work hardening, in real materials the material near the surface is strain hardened more than the interior, which in turn causes the stresses at the surface to be more tensile than predicted by theory.

Any residual stresses in the product are generally undesirable because they lower the elastic limit of the product and cause a tendency to warpage during subsequent machining operations. The usual residual stress patterns with surface tension are particularly undesirable, for they cause an increased susceptibility to fatigue and stress corrosion.

10-7 PLANE-STRAIN VERSUS AXISYMMETRIC DEFORMATION

Friction, redundant strain, and inhomogeneity effects on plane-strain and axisymmetric deformation are summarized and compared in Table 10-1. General conclusions indicate that frictional effects are primarily a function of die angle. Equal frictional effects on plane strain and axisymmetry should be found at the same values of α and r. If comparison is made in terms of Δ, an axisymmetric process should be compared with a plane-strain operation with half as large a value of Δ.

TABLE 10-1 Comparison of Axisymmetric and Plane-Strain Drawing (Extrusion)

	Plane Strain	Axisymmetry	Comment
A. Ratio of tool-workpiece contact area to mean cross-sectional area.	$\dfrac{2r}{(2-r)\sin\alpha} = \dfrac{2}{\Delta}$	$\dfrac{2r}{(2-r)\sin\alpha} \simeq \dfrac{4}{\Delta}$	equal for same r, α but with same Δ, axisymmetry has twice the contact area
B. Frictional contribution to $\sigma_d/2k\epsilon$ in slab analysis with constant shear stress, mk.	$\dfrac{m}{\sin 2\alpha}$	$\dfrac{m}{\sin 2\alpha}$	equal for same Δ and r, axisymmetry has twice the friction effect
C. Frictional contribution to $\sigma_d/2k\epsilon$ in slab analysis with constant coefficient of friction.	$\approx \mu \cot\alpha\left(1 - \dfrac{\epsilon}{2}\right)$	$\approx \mu \cot\alpha\left(1 - \dfrac{\epsilon}{2}\right)$	equal for same r, α but with same Δ, axisymmetry has about twice the friction effect
D. Redundant strain contribution to $\sigma_d/2k\epsilon$ according to upper-bound slab analysis.	$\dfrac{\tan\alpha}{2\epsilon} \approx \dfrac{r\Delta}{2(2-r)\epsilon}$ $\approx \dfrac{\Delta}{4}$	$\dfrac{2}{3}\dfrac{\tan\alpha}{\epsilon} \approx \dfrac{r\Delta}{3(2-r)\epsilon}$ $\approx \dfrac{\Delta}{6}$	for same r, α less redundant work in plane strain but for same Δ, less in axisymmetry
E. Redundant strain according to slip-line field for indentation.	$\approx 0.23(\Delta - 1)$	no solution for axisymmetry	
F. Redundant strain; from Burke and Backofen experiments on copper.†	$0.21(\Delta - 1)$	$0.12(\Delta - 1)$	for same Δ nearly twice as much redundant strain in plane strain; for same r and α more redundant strain in axisymmetry
G. Inhomogeneity Factor $(H_s - H_c)/H_c$ from Burke and Backofen experiments.†	primarily a function of α		approximately equal for axisymmetry and plane strain at same α

†W. A. Backofen, *ibid.*

Similarly, at least for low values of α, inhomogeneity in both operations appears to have a similar and primary dependence upon α. At larger reductions, the data of Miura et al.† indicate a Δ-dependence for axisymmetric deformation. Redundant strain correlates best with Δ. For equal values of Δ, the plane-strain operations are characterized by much higher ϕ-values than axisymmetric deformation, but if comparison is made at equal tool-workpiece to mean cross-sectional areas, the values of ϕ are somewhat higher in axisymmetry.

PROBLEMS

10-1. The XYZ Special Alloy Fabrication, Inc. has a business of supplying small lots (up to 500 lb) of special alloys fabricated to order. One job called for a slab of a nickel-base superalloy, 4 in. wide and $\frac{1}{2}$ in. thick. They cast a standard $4 \times 4 \times 15$-in. ingot and hot rolled it in their 12-in. diameter, 2-high mill, making reductions of about 5% per pass. Unfortunately, on the fifth pass the slab split longitudinally, parallel to the rolling plane. The project engineer, the shop foreman, and a consultant met to discuss how to proceed on a second ingot.

The project engineer proposed that forward and back tension be applied during rolling. The consultant suggested reducing the reduction per pass, while the shop foreman was in favor of heavier reduction per pass.

With whom, if anyone, do you agree? Explain your reasoning.

10-2. The Mannesmann process for making tubes from cylindrical billets is illustrated, and involves passing a billet between two rolls, whose axes are not parallel. The misalignment of the rolls causes the billet to spin as it is driven forward onto a mandrel, which pierces the billet and forms the tube. The head of the mandrel is positioned in the middle of the roll gap which is adjusted for a very light potential reduction.

Explain why the axial force on the mandrel is low, and why the mandrel, which is long and therefore elastically flexible, follows the center of the billet.

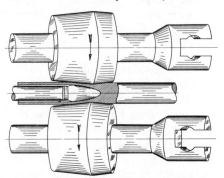

10-3. (a) Show that as r and α approach zero, Eqs. (10-11a and b) reduce to $\phi = 1 + (\Delta/4)$ for plane strain and $\phi = 1 + (\Delta/6)$ for axisymmetry.

(b) What is the percentage error in these approximations at $\alpha = 30°$, $r = 0.5$?

†Miura et al., *ibid.*

10-4. Consider wire drawing with a reduction of $r = 0.25$ in a die of semi-angle $\alpha = 6°$.

    (a) Calculate the ratio of the contact area between tool and workpiece to the average cross-sectional area of the deforming zone.

    (b) What is the value of Δ?

    (c) For strip drawing with $\alpha = 6°$, what reduction would give the same value of Δ as in part b?

    (d) For the reduction in part c, calculate the ratio of contact area between the tool and workpiece to the cross-sectional area in the middle of the die and compare with your answer to part a.

    (e) Explain why for the same Δ the frictional drag is greater in axisymmetric drawing.

10-5. Some authorities have defined Δ as the ratio of contact length to the length of an arc across the mid-section of the deformation zone, this arc being centered on the apex of a cone or wedge formed by extrapolating the die walls.

    (a) Show that for wire drawing this definition corresponds to $\Delta = \alpha(1 + \sqrt{1 - r})^2/r$.

    (b) Calculate the ratio of Δ from this definition to the value of Δ from Eq. (10-4) for

       i) $\alpha = 10°, r = 0.25$

       ii) $\alpha = 10°, r = 0.5$

       iii) $\alpha = 45°, r = 0.5$

       iv) $\alpha = 90°, r = 0.5$

10-6. Backofen developed the following equation for the mechanical efficiency, η, during wire drawing.

$$\eta = \left[1 + C(\Delta - 1) + \frac{\mu}{\tan \alpha}\right]^{-1} \quad \text{for } \Delta > 1$$

where C is an empirical constant equal to about 0.12, μ is the friction coefficient, α is the semi-die angle, and Δ is the ratio of the mean diameter of the wire in the die gap to the contact length between the die and the wire. Note the term $C(\Delta - 1)$ represents w_r/w_i and the term $\mu/\tan \alpha$ represents w_f/w_i so

$$\eta = \left(\frac{w_i}{w_i} + \frac{w_r}{w_i} + \frac{w_f}{w_i}\right)^{-1} = \left(\frac{w_a}{w_i}\right)^{-1}$$

    (a) Using Backofen's equation with $\mu = 0.05$ and $r = 0.3$, calculate the values of η for semi-die angles of $\alpha = 2, 4, 6, 8, 10, 12, 14,$ and $16°$ and plot η versus α. What is the optimum die angle, α^*, for these conditions?

    (b) Derive an expression showing how the optimum semi-die angle, α^*, varies with μ and r. Does α^* increase or decrease with μ? with r?

10-7. (a) Referring to Secs. 8-11 and 8-12, derive an expression for the redundant strain in drawing at the surface and at the centerline. (*Hint:* the redundant strain arises from the shear discontinuity at the die entrance and die exit, and for a non-work hardening material $\simeq w_r/\sigma_o$.)

    (b) Using the result of part a, find an expression for the ratio of the total strain at the surface to that at the centerline, ϵ_s/ϵ_c.

    (c) It has been suggested that the Vickers hardness for a cold-worked material is given by

$$H = C\epsilon^n$$

where n is not necessarily equal to the strain-hardening exponent in a tension test. Using this equation and the results from part b, derive an expression for the inhomogeneity factor, I.F., in terms of α and the homogeneous strain.

(d) The reference on which Fig. 10-7 is based gives $H = 134\epsilon^{0.15}$. Using the result of part c and $n = 0.15$, make a plot of I.F. versus α for $r = 0.08, 0.16,$ and 0.32 for $0 \leq \alpha \leq 30°$ and compare with Fig. 10-7.

10-8. A slab of annealed copper, with dimensions $x = 1$ in., $y = 1\frac{1}{2}$ in., and $z = 5$ in., is to be compressed between flat, rough, non-lubricated platens until the x dimension is reduced to 0.90 in.

Describe as fully as possible any inhomogeneity of resulting hardness. Where on the slab would the amount of work hardening be the least and where would the most work hardening occur?

10-9. A high strength steel bar must be cold reduced from a diameter of 1.00 to 0.65 in. by drawing. A number of schedules has been proposed. Which schedule would you choose to avoid drawing failure and minimize the likelihood of centerline bursts? Assume $\eta = 0.50$. Show your reasoning.

(a) A single reduction in a die of semi-angle 8°.

(b) Two passes (1.00 to 0.81 and 0.81 to 0.65 in. dia.) through dies of semi-angle 8°.

(c) Three passes (1.00 to 0.87, 0.87 to 0.75, and 0.75 to 0.65 in. dia.) through dies of $\alpha = 8°$.

(d) Four passes (1.00 to 0.90, 0.90 to 0.81, 0.81 to 0.72, 0.72 to 0.65 in. dia.) through dies of semi-angle 8°.

(e, f, g, and h) Same reduction schedules as a, b, c, and d respectively except using dies of $\alpha = 15°$.

11

Formability

11-1 INTRODUCTION

An important concern in forming is whether the desired deformation can be accomplished without failure of the work metal. For a given process and deformation geometry, the forming limits vary from material to material. As indicated in Chap. 10, central bursts may occur in extrusion at certain levels of Δ, yet not all materials will fail under these conditions. Also, the failure strains for a given material depend upon the process.

11-2 DUCTILITY

In most bulk forming operations, formability is limited by ductile fracture.† The forming limits in many processes correlate quite well with the reduction in area measured in a tension test. Fig. 11-1 shows the strains at which edge cracks first occur in rolling as a function of the tensile reduction in area expressed as a fracture strain, ϵ_f. The fact that the limiting strains for square-edged strips are much higher than those for round-edged strips indicates that process variables are also important. See end of Sec. 7-12.

---

†Wire drawing is an exception; the maximum reduction per pass is limited by the ability of the drawn section to carry the required drawing force without yielding and necking. Once the drawn wire necks, it can no longer support the required load, so the subsequent fracture strain is of little concern. It is also possible, though not common, that brittle fracture limits formability.

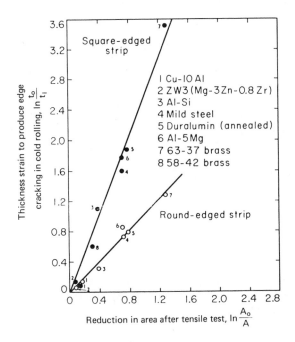

Figure 11-1 Correlation of strain at which edge cracking occurs in flat rolling with reduction of area in tension test. From M. G. Cockcroft and D. J. Latham, *J. Inst. Metals*, 96 (1968), pp. 33–39.

Similar results are reported for other processes. This correlation of limiting strains in forming with tensile ductility is perhaps surprising because ductile failures in tension tests start in the center of the bar, whereas cracks in most forming operations initiate at the surface.

11-3 METALLURGY

The ductility of a metal is strongly influenced by the properties of both the matrix and inclusions. In general, factors which increase the strength of the matrix tend to reduce ductility. Solid-solution strengthening, precipitation, cold work, and decreased temperatures all lower fracture strains. The reason is that with increased flow strengths, higher stresses will be encountered in tension testing and in forming.

Inclusions play a dominant role in ductile fracture. The volume fraction, nature, shape, and distribution of inclusions are important. In Fig. 11-2, the tensile ductility of copper is seen to decrease with increased amounts of artificially added inclusions. Despite the scatter it appears that some second-phase particles are more deleterious than others. Figure 11-3 shows similar trends for oxides, sulfides and carbides in steels.

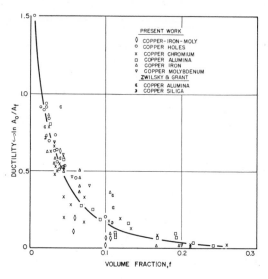

Figure 11-2 Effect of volume fraction of second-phase particles in copper on tensile ductility. From B. I. Edelson and W. M. Baldwin, *Trans. Q. ASM*, 55 (1962), pp. 230–50.

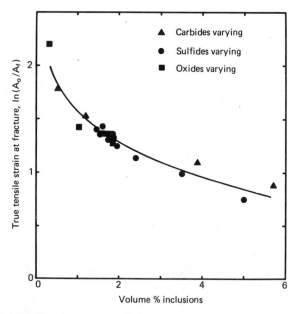

Figure 11-3 Effect of volume fraction natural inclusions on tensile ductility of steel. Adapted from F. B. Pickering, *Physical Metallurgy and Design of Steels* (London: Applied Science Pub., 1978), p. 51.

Mechanical working tends to elongate and align inclusions in the direction of extension. This mechanical fibering reduces the fracture strength and ductility normal to the principal working direction. Wrought iron, though no longer a commercial product, is an extreme example. Crack paths follow glassy silicate inclusions producing "woody" fractures (see Fig. 11-4). In steels, MnS inclusions elongated during hot working are a major cause of anisotropic

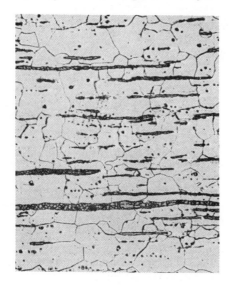

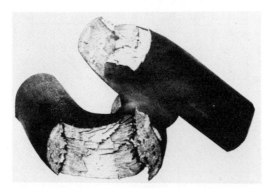

Figure 11-4   Microstructure of wrought iron showing elongated glassy silicate inclusions (100x, rolling direction horizontal) (top) and the woody fracture of wrought iron caused by cracking along the inclusions (bottom). From J. Aston and E. B. Story, *Wrought Iron and its Manufacture, Characteristics and Applications* (Pittsburgh: Byer Co., 1939), pp. 45, 92.

fracture behavior as illustrated in Fig. 11-5. Transverse ductility of steels can be markedly improved by *inclusion shape control*, which is the practice of adding small amounts of calcium, cerium, titanium, or rare earths (see Fig. 11-6). These react with the sulfur to form hard inclusions that remain spherical

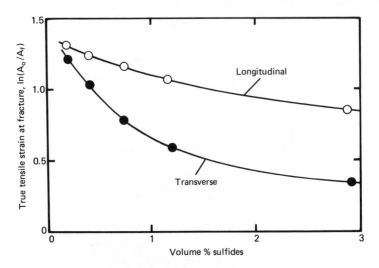

Figure 11-5 Effect of sulfides on tensile ductility of steel. The lower transverse ductility is a result of inclusions being elongated in the rolling direction. Adapted from F. B. Pickering, *ibid.*, p. 82.

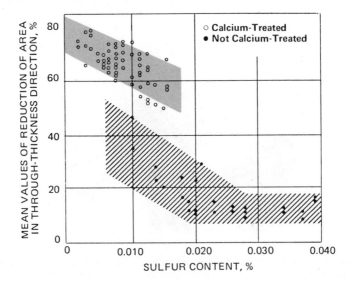

Figure 11-6 Effect of inclusion shape control in raising the through-thickness ductility of an HSLA steel. From H. Pircher and W. Klapner, in *Micro Alloying 75*, Union Carbide, N.Y.: (1977), pp. 232–40.

during working. If the added cost is justified, inclusion content can be greatly reduced by desulfurization or vacuum remelting.

Scanning electron microscope pictures of ductile fracture surfaces (e.g., Fig. 11-7) are characterized by dimples usually associated with inclusions. These observations suggest that the fracture initiates by the opening of holes or voids around inclusions; these voids grow by plastic deformation until they link up to form macroscopic cracks. Figure 11-8 shows a crack near the center of the neck in a tensile bar of cold-rolled 1018 steel prior to final fracture. From such observations, the role of inclusions is clear.

The joining of voids may occur by their preferential growth parallel to the axis of highest principal stress until their length, λ, approximately equals

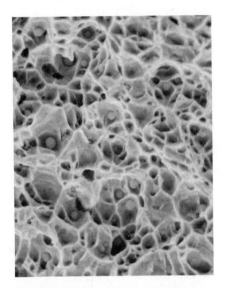

Figure 11-7 Dimpled ductile fracture surface of steel. Note that the microvoids are associated with inclusions (3000x, reduced 6% for publication). Courtesy of J. W. Jones.

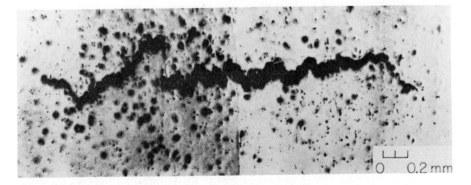

Figure 11-8   Crack in tensile specimen of cold-rolled 1018 steel. Oil stain from diamond polish at left exaggerates the hole size. From F. McClintock, in *Ductility*, ASM, Metals Park, Ohio (1968), pp. 255–77.

their separation distance, *d*. The ligaments left between the voids then undergo necking on a microscopic scale, as suggested in Fig. 11-9.

Ductile fracture may also occur by localized shear. One example is the shear lip on a ductile cup-and-cone fracture in a tension test. Shear failures may also occur in rolling and in other forming processes. Void formation around inclusions also plays a dominant role in shear fracture, as indicated in Figs. 11-10 and 11-11.

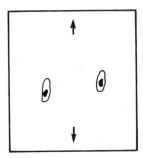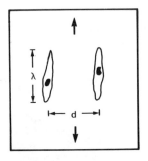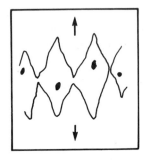

Figure 11-9 Schematic illustration of how a ductile crack can form under tension by formation of elongated voids at inclusions, and the necking of the ligaments between voids.

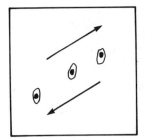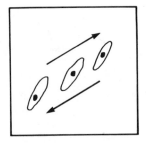

Figure 11-10 Schematic illustration of how a ductile crack can form in shear by linking up of voids formed around inclusions.

11-4 HYDROSTATIC STRESS

Fracture strains depend upon the level of hydrostatic stress during deformation as shown in Fig. 11-12. High hydrostatic pressure suppresses void growth, thereby delaying fracture. Conversely, hydrostatic tension accelerates void growth and decreases ductility. One measure of the extent of void growth is the decrease in bulk density discussed in Chap. 10. Figure 11-13 shows the effect of hydrostatic pressure on density changes during strip drawing of 6061-T6 aluminum. The dependence of fracture strain on process geometry, as indicated in Fig. 11-1, can also be explained by differing levels of hydrostatic stress.

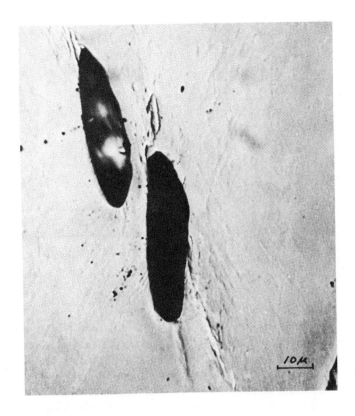

Figure 11-11 Photomicrograph of localized shear with large voids in OFHC copper. From H. C. Rogers, *Trans. TMS-AIME*, 218 (1960), pp. 498–506.

Figure 11-12 Correlation of fracture strain with ratio of hydrostatic stress to effective stress. From A. L. Hoffmanner, *Interim Report, Air Force Contract F 33615-67-C-1466*, TRW (1967); see also, G. E. Dieter, in *Ductility*, ASM (1968), pp. 1–30.

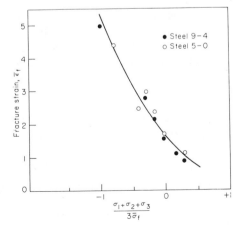

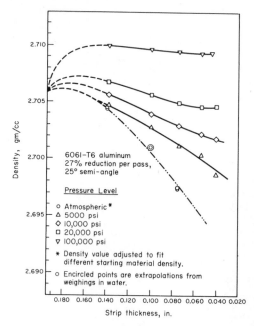

6061–T6 aluminum
27% reduction per pass,
25° semi-angle

Pressure Level

○ Atmospheric *
△ 5000 psi
◇ 10,000 psi
☐ 20,000 psi
▽ 100,000 psi

* Density value adjusted to fit
different starting material density.

○ Encircled points are extrapolations from
weighings in water.

Density, gm/cc

2.710
2.705
2.700
2.695
2.690

0.180 0.160 0.140 0.120 0.100 0.080 0.060 0.040 0.020

Strip thickness, in.

Figure 11-13 Loss of density by growth of microporosity during strip drawing, and the effect of superimposed hydrostatic pressure on diminishing the density loss. Note that drawing at the highest pressure level increased the density, presumably by closing up preexisting pores formed during earlier processing. From H. C. Rogers and L. F. Coffin, Jr., *Final Rept., Contract NOw-66-0546-d* (June 1967), Naval Air Systems Command; see also, H. C. Rogers, in *Ductility*, ASM (1968), pp. 31–61.

If all stress components could be maintained compressive, formability problems would be minimized. Materials of very limited formability can be successfully extruded if both the billet and die exit region are under high hydrostatic pressure. Without such special conditions, tensile stresses are apt to result in processes which are nominally compressive. Examples mentioned previously include the edges of strips during rolling and the central region during extrusion with high Δ. During forging and upsetting, barrelling causes tensile hoop stresses at the free surface. These occur because the axial compressive stress is lower at the free surface than elsewhere and the interior expands radially, causing hoop tension at the surface.

Although some workers have suggested that the failure strain depends upon the level of the largest principal stress, σ_1, rather than on the hydrostatic stress, σ_m, this distinction is not very important, as can be seen by comparing extreme possibilities. For axisymmetric elongations ($\epsilon_2 = \epsilon_3 = -\frac{1}{2}\epsilon_1$), $\sigma_m = \sigma_1 - \frac{2}{3}\bar{\sigma}$, while for axisymmetric compression ($\epsilon_1 = \epsilon_2 = -\frac{1}{2}\epsilon_3$), $\sigma_m = \sigma_1 - \frac{1}{3}\bar{\sigma}$. All other flow conditions are intermediate so

$$\sigma_m + \frac{1}{3}\bar{\sigma} \leq \sigma_1 \leq \sigma_m + \frac{2}{3}\bar{\sigma} \tag{11-1}$$

Thus, σ_m and σ_1 usually increase or decrease together.

Cockcroft and Latham† have suggested a fracture criterion of:

$$\int_0^{\epsilon_f} \sigma_1 \, d\bar{\epsilon} = C \tag{11-2}$$

The concept is that damage accumulates during processing until a critical value,

†M. G. Cockcroft and D. J. Latham, *J. Inst. Metals*, 96 (1968), pp. 33–39.

C, is reached at ϵ_f. This criterion recognizes the dependence of fracture strain upon σ_1 (or almost equivalently, σ_m) and on the loading path. However, if σ_1 is even temporarily negative, it predicts accumulation of negative damage, a concept that is difficult to justify. For this and other reasons, many other fracture criteria have been proposed.[†]

11-5 BULK FORMABILITY TESTS

Formability does not always correlate well with tensile ductility, particularly when cracking initiates at a free surface. Furthermore, the size and shape of the work metal often preclude making tension tests in the relevant directions. For these reasons other formability tests may be used.

Kuhn[‡] has proposed a simple procedure to evaluate formability using cylindrical upset tests. Barrelling during compression induces tensile hoop stresses on the surface that eventually cause cracking. The amount of barrelling and hence the circumferential and axial strains, ϵ_{1f} and ϵ_{2f}, at fracture can be altered by changing the lubrication and the height-to-diameter ratio of the specimens. A plot of ϵ_{1f} versus ϵ_{2f} from a series of specimens is shown in Fig.

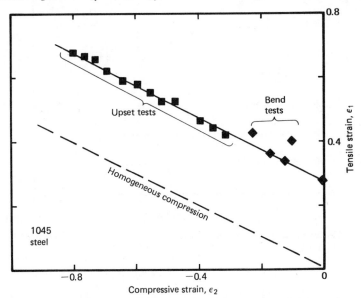

Figure 11-14 Forming limits defined by strains at fracture in upset and formability tests. Adapted from H. A. Kuhn, *Formability Topics-Metallic Materials*, ASTM STP 647 (1978), pp. 206–19.

[†]See, for example: A. G. Atkins, *Metal Sci.* (Feb., 1981), pp. 81–83; A. K. Ghosh, *Met. Trans.*, 7A (1976), pp. 523–33; F. A. McClintock, *J. Appl. Mech., Trans. ASME*, 35 (1968), pp. 363–71.

[‡]H. A. Kuhn, *Formability Topics-Metallic Materials*, ASTM STP 647 (1978).

11-14. Additional points near plane-strain conditions ($\epsilon_2 = 0$) were obtained by measuring fracture strains in the bending of wide specimens. The straight line indicates that

$$\epsilon_{1f} = C - \frac{1}{2}\epsilon_{2f} \qquad (11\text{-}3)$$

where C is the value of ϵ_{1f} for plane strain ($\epsilon_2 = 0$). This line parallels the line $\epsilon_1 = -\frac{1}{2}\epsilon_2$ for homogeneous compression without barrelling, so ductile fracture under homogeneous compression would not be expected. These findings suggest that a single test would be sufficient to evaluate C for a material and, therefore, the formability under a variety of strain paths. Results obtained with other specimen orientations indicate the fracture anisotropy as shown in Fig. 11-15.

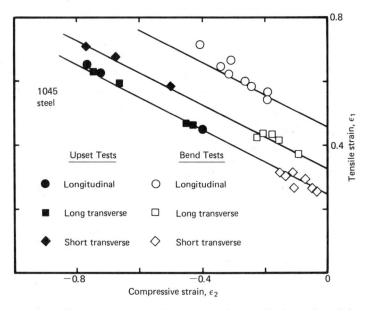

Fig. 11-15 Effect of specimen orientation on forming limits. Adapted from H. A. Kuhn, *ibid.*

11-6 HOT WORKING

Formability in hot working differs from that in cold working. Usually, hot formability is much greater than cold formability, but sometimes failure occurs at very low strains. This is called *hot shortness*. The usual cause is the presence of a liquid phase, which may occur because of partial melting caused by heating into a two-phase, liquid-solid region. The hot-working temperatures for many aluminum alloys are very close to the eutectic temperature; in other cases the liquid may be a non-metallic phase. In the absence of manganese, sulfur would be very detrimental in steels because FeS is molten at normal hot-working temperatures and forms a continuous or semicontinuous film along the grain boundaries. Fortunately this condition is easily prevented by additions of manganese, which combine readily with sulfur. Manganese sulfide is solid at hot-working temperatures and does not wet the grain boundaries. Small amounts of copper or tin, picked up from scrap, can cause surface cracking during hot rolling of steel. These "tramp" elements do not oxidize in the presence of iron, so as the surface of the billet is oxidized during heating prior to rolling, they are concentrated in solid solution in the region just beneath the oxide scale. If the local concentration exceeds the solid solubility, a separate liquid phase will form; this wets the grain boundaries and causes a loss of ductility.

Inclusions also affect hot formability, acting in a similar way to that in cold forming. Highly deformable inclusions (e.g., MnS) are elongated in the working direction. This is not a serious problem if working is continued in the same direction, but if the working direction is altered, as in hot upsetting or heading of hot-rolled rod, the effects can be severe. Here again, inclusion shape control would be beneficial.

PROBLEMS

11-1. (a) Explain why inclusion-shape control is of much greater importance in high-strength steels than in low-carbon steels.

(b) Explain why inclusion-shape control increases the transverse ductility and toughness but has little effect on the longitudinal properties.

11-2. Figure 11-1 shows that, for a given sheet material, greater reduction is possible before edge cracking if square edges are maintained than if the edges are rounded. Explain this observation in terms of the stress state at the edge of the strip.

11-3. (a) Wrought iron has a very high toughness when stressed in tension parallel to the prior rolling direction, but a low toughness when stressed perpendicular to the prior rolling direction. Explain why.

(b) When wrought iron was a commercial product, producers claimed that wrought iron sheet and pipe were more corrosion resistant than comparable products of low-carbon steel. What was the basis, if any, of their claim? Explain.

11-4. Figure 11-13 shows that the density of 6061-T6 aluminum can be increased by drawing at very high levels of external pressure. How is this possible?

11-5. (a) With the same material, die angle, and reduction, central bursts may be encountered in drawing but not in extrusion. Explain.

(b) A given material may be rolled to a higher reduction of area before fracture than that observed in a tension test. Explain.

Bending

12-1 INTRODUCTION

All sheet-forming operations incorporate some bending; often it is the major feature, and several aspects of bending warrant consideration. After forming, some elastic springback occurs and considerable residual stresses may result. If the bend radius is too sharp, excessive tensile strain on the outside surface may cause failure. Sometimes buckling, due to excessive compressive strain, is encountered on the inside of a bend. To develop an understanding of the mechanics of bending, a very simple case will be treated initially, this being the bending of a flat sheet of a non-work hardening material subjected to a pure bending moment.

The coordinate system is shown in Fig. 12-1. Let r be the radius of curvature measured to the mid-plane and z be the distance of any element from the mid-plane. The engineering strain at z can be derived by considering arc lengths, L, measured parallel to the x-axis. The arc length at the midplane, L_o, doesn't change during bending, and may be expressed as $L_o = r\theta$, where θ is the bend angle. At z the arc length is $L = (r + z)\theta$. Before bending, both lengths were the same, so the engineering strain is $e_x = (L - L_o)/L_o = z\theta/r\theta = z/r$. The true strain is

$$\epsilon_x = \ln\left(1 + \frac{z}{r}\right) \tag{12-1a}$$

but for many bends the strains are small enough that we can take

$$\epsilon_x \approx \frac{z}{r} \tag{12-1b}$$

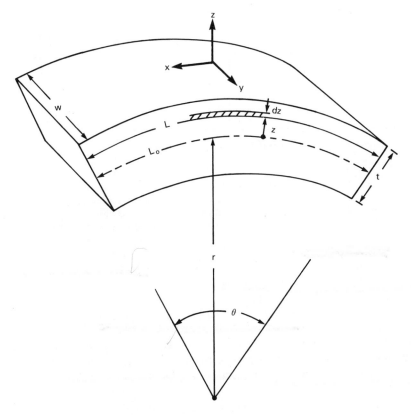

Figure 12-1 Coordinate system for analysis of bending.

12-2 SPRINGBACK IN SHEET BENDING

With sheet material, $w \gg t$, so width changes are negligible. Therefore, we can consider bending as approaching a plane-strain operation, $\epsilon_y = 0$, $\epsilon_z = -\epsilon_x$. The value of ϵ_x varies linearly from $-t/2r$ at the inside ($z = -t/2$), to zero at the mid-plane ($z = 0$), to $+t/2r$ at the outside ($z = t/2$) from Eq. 12-1b and Fig. 12-2a. Knowing the strain distribution, the internal stress distribution can be found if the slope of the stress-strain curve is known. First, assume a material that is elastic-ideally plastic (Fig. 12-2b). If the tensile yield stress is Y, the flow stress in plane strain will be $\sigma_o = \sqrt{4/3}\,Y$. Figure 12-2c shows the stress distribution through the sheet. The entire section will be at a stress, $\sigma_x = \pm\sigma_o$, except for an elastic core near the mid-plane, which will shrink as the bend radius decreases. For most bends this elastic core can be neglected.

To calculate the bending moment, M, needed to produce this bend, it is assumed that there is no *net* external force in the x direction ($\Sigma F_x = 0$). However, the internal force, dF_x, acting on any incremental element of cross section,

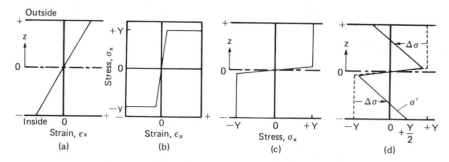

Figure 12-2 Strain and stress distribution across sheet thickness. Bending strain (a) varies linearly across the section. For the non-work hardening stress-strain relation (b), the bending moment causes the stress distribution in (c). Elastic unloading after removal of the loads results in the residual stresses shown in (d).

wdz, is $dF_x = \sigma_x wdz$. The contribution of this element to the bending moment is the product of the force and lever arm, z, so $dM = zdF_x = z\sigma_x wdz$. The total bending moment is found from

$$M = \int_{-t/2}^{+t/2} w\sigma_x zdz = 2\int_0^{t/2} w\sigma_x zdz \qquad (12\text{-}2)$$

(With integration limits of $-t/2$ to $t/2$, the sign of both σ_x and z change at the mid-plane, so it is simpler to double the value of the integral from 0 to $t/2$.) For the ideally plastic material with a negligible elastic core, $\sigma_x = \sigma_o$, and

$$M = 2w\sigma_o \int_0^{t/2} zdz = w\sigma_o \frac{t^2}{4} \qquad (12\text{-}3)$$

Example 12-1

A steel sheet, 0.036 in. thick, is bent to a radius of curvature of 5.0 in. The flow stress is 33×10^3 psi.
a) What fraction of the cross section remains elastic?
b) What percent error does neglecting the elastic core cause in the calculation of the bending moment?

Solution

a) The elastic strain at yielding is $\epsilon_x = \sigma_o/E'$, where E' is the plane-strain modulus, $E/(1 - \nu^2)$. Using Eq. (12-1b), the limit of the elastic core will be at $z = r\epsilon_x = r\sigma_o/E'$. Taking E' as 33×10^6 psi, $z = 5 \times 33 \times 10^3/33 \times 10^6 = 0.005$ in. The elastic fraction is $2 \times 0.005/0.036 = 0.28$ or 28%.
b) To calculate the bending moment, divide the integral in Eq. (12-2) into two parts; for the elastic portion $(0 \leq z \leq 0.005)$, $\sigma_x = \epsilon_x E' = zE'/r$, and for the plastic portion $(0.005 \leq z \leq 0.018)$, $\sigma_x = \sigma_o$.

$$M = 2\int_0^{0.005} w\frac{E'}{r}z^2\,dz + 2\int_{0.005}^{0.018} w\sigma_o zdz$$

$$= \frac{2}{3}\frac{33 \times 10^6}{5}(0.005)^3 w + 33 \times 10^3(0.018^2 - 0.005^2)w = 10.42w$$

Using Eq. 12-3, which neglects the elastic core,

$$M = (33 \times 10^3)\frac{(0.036^2)w}{4} = 10.96w$$

The error is $(10.69 - 10.42)/10.42 = 0.026$ or 2.6%.

The external moment applied by the tools and the internal moment resisting bending must be equal, so Eq. (12-3) applies to both. When the external moment is released, the internal moment must also vanish. As the material unbends (springs back) elastically, the internal stress distribution results in a zero bending moment. Since the unloading is elastic,

$$\Delta\sigma_x = E'\Delta\epsilon_x \tag{12-4}$$

where, because of plane strain, $E' = E/(1 - v^2)$. The change in strain is given by

$$\Delta\epsilon = \frac{z}{r} - \frac{z}{r'} \tag{12-5}$$

where r' is the radius of curvature after springback. This causes a change in bending moment, ΔM, of

$$\Delta M = 2w \int_o^{t/2} \Delta\sigma_x z\,dz = 2w \int_o^{t/2} E'\left(\frac{1}{r} - \frac{1}{r'}\right)z^2\,dz$$

$$\Delta M = \frac{wE't^3}{12}\left(\frac{1}{r} - \frac{1}{r'}\right) \tag{12-6}$$

Since $M - \Delta M = 0$ after springback, equating Eqs. (12-3) and (12-6) gives

$$\frac{wE't^3}{12}\left(\frac{1}{r} - \frac{1}{r'}\right) = \frac{w\sigma_o t^2}{4}$$

or

$$\frac{1}{r} - \frac{1}{r'} = \frac{3\sigma_o}{tE'} \tag{12-7}$$

The resulting residual stress, $\sigma'_x = \sigma_x - \Delta\sigma_x = \sigma_o - E'\Delta\epsilon_x = \sigma_o - E'z(1/r - 1/r') = \sigma_o - E'z(3\sigma_o/tE')$,

$$\sigma'_x = \sigma_o\left(1 - \frac{3z}{t}\right) \tag{12-8}$$

This is plotted in Fig. 12-2d. Note that on the outside surface where $z = t/2$, the residual stress is compressive, $\sigma'_x = -\sigma_o/2$, and on the inside surface ($z = -t/2$) it is tensile, $\sigma'_x = +\sigma_o/2$.

A similar development can be made for a work-hardening material. If $\bar\sigma = K\bar\epsilon^n$, then $\sigma_x = K'\epsilon_x^n = K'(z/r)^n$, where $K' = K(\frac{4}{3})^{(n+1)/2}$. Substituting σ_x into Eq. (12-2) and continuing the analysis,

$$M = \left(\frac{2}{2+n}\right)\frac{wK'}{r^n}\left(\frac{t}{2}\right)^{2+n} \tag{12-9}$$

Since ΔM is still described by Eq. 12-6 and $M - \Delta M = 0$ after springback,

$$\left(\frac{1}{r} - \frac{1}{r'}\right) = \left(\frac{6}{2+n}\right)\left(\frac{K'}{E'}\right)\left(\frac{t}{2r}\right)^n\left(\frac{1}{t}\right) \qquad (12\text{-}10)$$

Using Eqs. (12-4), (12-5), and (12-10) gives

$$\sigma_x' = K'\left(\frac{z}{r}\right)^n\left[1 - \left(\frac{3}{2+n}\right)\left(\frac{2z}{t}\right)^{1-n}\right] \qquad (12\text{-}11)$$

The variations of σ_x, $\Delta\sigma_x$, and σ_x', through the section are shown in Fig. 12-3. The magnitude of the springback predicted by Eqs. (12-7) or (12-10) can be very large.

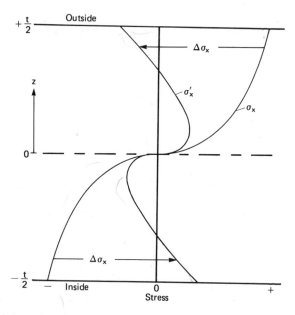

Figure 12-3 Stress distribution under bending moment and after unloading for a work-hardening material.

Example 12-2

Find the tool radius necessary to produce a final bend radius of $r' = 10$ in. in a part made from a steel of thickness 0.030 in. Assume a flow stress of 45,000 psi.

Solution. Substituting $r' = 10$, $t = 0.030$, $\sigma_o = 45 \times 10^3$, and $E' = 33 \times 10^6$ into Eq. (12-7), $r = 4.2$ in. If the bend were in a portion of a complex stamping it would be almost impossible to design tooling for so much of an overbend. (Actually the springback problem is somewhat greater, since at a bend radius of 4.2 in., the limit of the elastic core is at $z = r\sigma_o/E' = (4.2)(45 \times 10^3)/(33 \times 10^6) = 0.0057$ in. which is 38% of the cross section. This causes an error of 5.1% in the calculation of the moment. The problem would be more severe with larger radii of curvature.

12-3 BENDING WITH SUPERIMPOSED TENSION

Such allowances for springback would cause severe problems in tool design, but fortunately there is a relatively simple solution. Often, as in stretch forming, the tooling does not apply a pure bending moment as assumed above. Rather, tension is applied simultaneously with bending. With increasing tensile forces, F_x, the neutral plane shifts towards the inside of the bend and in most operations, this tension is sufficient to move the neutral plane completely out of the sheet so that the entire cross section yields in tension. For such a case, the strain and stress distributions are sketched in Fig. 12-4.

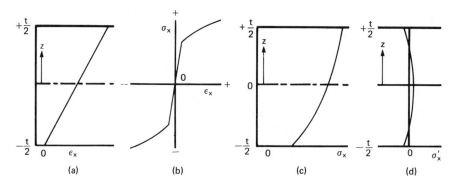

Figure 12-4 Bending with superimposed tension. With sufficient tension, the neutral axis moves out of the sheet so the strain is tensile across the entire section, (a). With the stress-strain curve shown in (b), the stress distribution in (c) results. After removal of the moment, elastic unloading leaves very minor residual stresses, as shown in (d).

In an ideally plastic material, there would be no induced bending moment and no springback because $\Delta\sigma_x = \sigma_o$ is constant across the section. With a work-hardening material the springback doesn't vanish with applied tension, but it is greatly reduced, as shown in Fig. 12-4d. It is convenient to consider the strain as consisting of two parts; a tensile strain, ϵ_t, which is constant across the section, and a bending strain, ϵ_b, which varies with distance from the centerline, $(\epsilon_b = z/r)$. To simplify the problem, the portion of the stress-strain curve in the range of loading will be approximated by a straight line of slope $d\sigma/d\epsilon$, so the bending stress, σ_b, is given by $\sigma_b = \epsilon_b(d\sigma/d\epsilon) = (z/r)(d\sigma/d\epsilon)$. Relaxation of the bending stress, σ_b, will cause a change of the bending strain $\Delta\epsilon_b = \sigma_b/E'$ $= (z/r)(d\sigma/d\epsilon)/E'$, leaving a final bending strain $\epsilon'_b = \epsilon_b - \Delta\epsilon_b = (z/r)$ $[1 - (d\sigma/d\epsilon)/E']$. Since the relaxed radius of curvature r' is given by $z/r' = \epsilon'_b$,

$$\frac{z}{r'} = \left(\frac{z}{r}\right)\left(1 - \frac{(d\sigma/d\epsilon)}{E'}\right)$$

or

$$\frac{r'}{r} = \left(1 - \frac{(d\sigma/d\epsilon)}{E'}\right)^{-1}$$

(12-12)

Springback is very much diminished from that predicted by Eq. (12-7), since the slope in the plastic range ($d\sigma/d\epsilon$) is very much less than E'.[†]

Example 12-3

Reconsider Ex. 12-2 with sufficient tension to cause a net tensile strain of 0.02 at the centerline. Let the stress-strain curve be approximated by $\sigma = 100,000\epsilon^{0.2}$, (this gives $\sigma = 46 \times 10^3$ psi at the centerline).

Solution. Substituting $d\sigma/d\epsilon = nK\epsilon^{n-1} = 0.2 \times 10^5(0.02)^{-0.8} = 457 \times 10^3$ and $E' = 33 \times 10^6$ into Eq. (12-12) gives $r'/r = 1.014$ or, with $r' = 10$ in., $r = 9.86$ in. Compare this with the value of $r = 4.2$ in. found with no net tension! Now it should be possible to design tools to compensate for springback.

12-4 SHEET BENDABILITY

During bending, the tensile strains on the outside of the bend may cause cracking. The limiting values of (r/t) vary with material; those materials with high reduction of area in a tension test tend to have low r/t limits, as shown in Fig. 12-5b. In this figure (and customarily in tables giving minimum bend radii) the radius, R, is measured to the *inside* of the bend rather than to the midplane as in this chapter. Therefore $R/t = r/t - 1/2$. The solid line represents the relation,

$$\frac{R}{t} = \frac{1}{2A_r} - 1 \qquad (12\text{-}13)$$

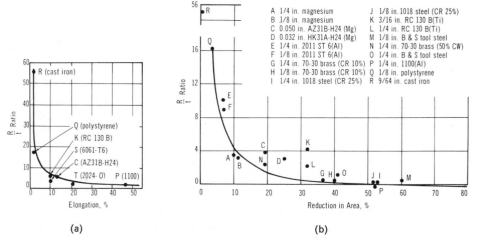

(a)   (b)

Figure 12-5 Correlation of limiting bend severity, R/t, with tensile ductility. Note that R is the inside radius of curvature. From C. T. Yang, *Metals Prog.* 98 (November 1970), pp. 107–10. See also, J. Datsko, *Materials Properties and Manufacturing Processes* (New York: John Wiley, 1966), p. 318.

[†]See J. L. Duncan and J. E. Bird, *Sheet Metal Industries* (September 1978), p. 1015.

where A_r is the decimal value of the reduction of area at fracture. This is derived by equating the bending strain in Eq. (12-1a) to the true strain at fracture, $\epsilon_f = \ln(A_o/A_f)$ in a tension test. While the correlation is reasonable for materials of limited ductility (high r/t), it is not accurate for sharp bends (low r/t) because the neutral axis shifts from the mid-plane and the amount of the shift depends upon the applied tension and the frictional conditions. Furthermore, the strains are not limited to the nominally bent region but spread into the adjacent nominally flat regions, so the strains predicted by Eq. (12-1b) are not reached at low bend angles. Despite the lack of quantitative agreement at low r/t, the permissible bend severity tends to increase with tensile ductility.

12-5 BENDING OF SHAPES AND TUBES

Consider now the bending of shapes (e.g., channels, "T"-, "L"-, and "I"-sections, square and round tubes, etc.), as shown in Fig. 12-6. Again, for simplicity we will assume no work hardening and pure bending without net x-direction force. With most shapes, it is better to assume that the stress in an element is x-direction uniaxial tension or compression, rather than plane-strain loading. The major change, compared with the earlier analysis, is that the section width, w, is a function of z and cannot be taken outside the integrals of Eqs. (12-2) and (12-6). For shapes symmetrical about the mid-plane (Fig. 12-6a and b),

$$M = 2\sigma_o \int_o^{h/2} wz\,dz$$

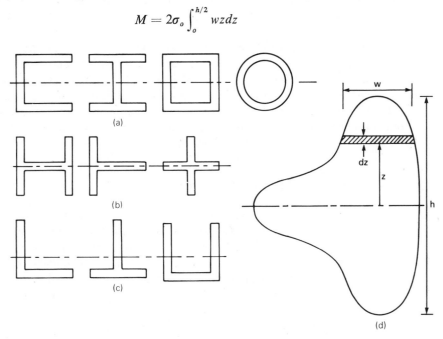

Figure 12-6   Cross sections of various shapes.

and
$$\Delta M = 2E\left(\frac{1}{r} - \frac{1}{r'}\right)\int_0^{h/2} wz^2 dz$$

so the relative springback is

$$\left(\frac{1}{r} - \frac{1}{r'}\right) = \frac{\sigma_o}{E} \cdot \frac{Q}{(h/2)} \qquad (12\text{-}14)$$

where
$$Q = \frac{h/2 \int_0^{h/2} wz\,dz}{\int_0^{h/2} wz^2 dz} \qquad (12\text{-}15)$$

and the residual stress is

$$\sigma_x' = \sigma_o\left(1 - \frac{2zQ}{h}\right) \qquad (12\text{-}16)$$

Note that for the flat sheet, $w \neq f(z)$, $Q = \frac{3}{2}$. For sections where w increases with z (Fig. 12-6a), $Q < \frac{3}{2}$, the springback relative to a flat sheet will be less for the same h, and the magnitude of the surface residual stresses will be lower. In sections where w decreases with z (Fig. 12-6b), $Q > \frac{3}{2}$ and both the springback and residual stresses will be greater than with a flat sheet. As in bending of flat sheet, springback can be diminished by applying tension during bending.

Up to this point, any shift of position in the neutral axis has been neglected. For symmetric bends and relatively large bend radii, any change in the position of the neutral axis during bending is small. With tight bends (small r/h), the neutral axis shifts toward the inside; there are several reasons for this. For shapes shown in Fig. 12-6, the cross section at the inside will increase while the outside decreases and the magnitude of the true strain (and hence the flow stress in a work-hardening material) increases faster with z in compression than tension.[†] As a consequence, the neutral axis moves inward to compensate for the higher stresses and greater cross section. In non-symmetric sections (Fig. 12-6c), transition from elastic to plastic flow will not occur simultaneously on both sides of the bend and, consequently, as yielding starts, there will be a shift of the neutral axis toward the heavier sections.

12-6 FORMING LIMITS IN SHAPE BENDING

In bending shapes of various cross sections, there are two potential modes of failure. Material on the outside of the bend can fail in tension by either necking or fracture, while compression on the inside of the bend can cause buckling.

In neglecting the neutral axis shift, the tensile strain on the outside of the bend is h/r, and failure occurs when this strain reaches a critical value, c; for successful bends, $h/r < c$. For brittle materials, c is approximately equal to the

---

[†]Recall that it is the engineering strain that varies linearly with z. The true strain on the tension side is less than the engineering strain while the reverse is true on the compression side.

percent elongation in a tension test, but for ductile materials the uniform percent elongation in a tension test is limited by necking. During bending, the outer fibers are supported by underlying material that suffers a lower strain and, therefore, is less prone to neck. The extent to which necking of extreme fibers is suppressed by the material closer to the neutral axis depends upon the section shape. In a sheet, necking can be completely suppressed, so failure must occur by fracture. For this case, critical bend radii correlate better with the reduction of area (A_r) in a tension test than with elongation, so $c \approx \ln(1 - A_r)^{-1}$. With thin-walled tubes there is little support from underlying fibers, so failure by necking is likely.

Buckling on the inside of the bend depends primarily on the slenderness of the section, h/t, but it also depends to some extent upon the bend severity, h/r. Wood† has determined forming limits for the bending of many materials and shapes in stretch forming (bending with superimposed tension). A schematic forming limit diagram is shown in Fig. 12-7. Note that with increasing tension, tensile failure occurs at lower values of h/r, while the critical h/t values for buckling increase.

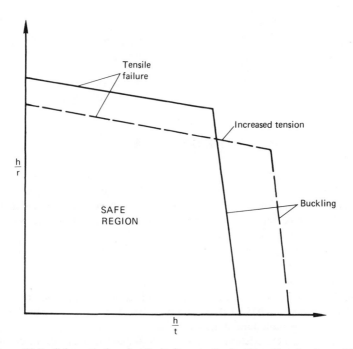

Figure 12-7   Schematic forming limit diagram for bending. Exact values of h/r and h/t depend upon the shape and material. Increasing axial tension lowers critical h/r and raises critical h/t.

†W. W. Wood, *Final Report on Sheet Metal Forming Technology*, Vol. II, ASD-TDR-63-7-871 (July, 1963).

When tubes are bent, there will be a distortion or even a collapse of the tubular section because outer fibers tend to move inward toward the neutral plane, where the tensile elongation is lower. To lessen this problem, the inside of the tube should be supported. With low production, the tube may be filled with sand or a low melting alloy, but with high production a ball or plug mandrel is often used. Internal support, while lessening the tendency of a section to collapse, will impose higher tensile strains on the outside, and consequently will lead to tensile failure at a lower bend severity, h/r. Figure 12-8 illustrates tube bending with a ball mandrel.

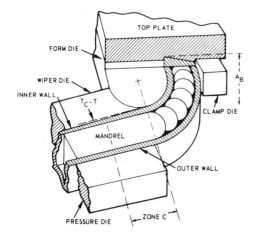

Figure 12-8 Bending of a square tube with ball mandrels to prevent collapse of tube walls. From *Cold Bending and Forming Tubes and Other Sections*, SME, Dearborn, Mich. (1966), p. 90.

PROBLEMS

12-1. An old shop hand has developed a method of estimating the yield strength of steel strips. He carefully bends the strip with his hands to a given radius of curvature, and then releases it and notes whether it fully returns to a flat shape. This procedure is repeated with ever-decreasing radii until the strip takes a permanent set when released. A strip, 0.025 in. thick, 0.75 in. wide, and 10 in. long, first takes a permanent set when bent to a radius of curvature of 10 in. Estimate its tensile yield strength.

12-2. A coiler is being designed for a cold-rolling line of a steel mill. The coil diameter should be large enough so that coiling involves only elastic bending. The sheet is 80 in. wide and 0.035 in. thick and has a yield strength of 40,000 psi.
   (a) What is the minimum diameter of the coiler?
   (b) For this diameter, determine the horsepower consumed in elastic deformation if the strip feeds into the coiler at 100 ft/sec.

12-3. In many design applications, the minimum sheet thickness is controlled by elastic stiffness requirements in bending. When aluminum is substituted for steel to save weight, the sheet thickness must be increased to compensate for its lower modulus. For flat sheets the elastic deflection δ is given by

$$\delta = A \frac{FL^3}{E' wt^3}$$

where F is the force, L the length of span, w the sheet width, t the sheet thickness, E' the plane-strain modulus, and A is a constant that depends upon the form of loading and nature of end supports.

(a) If aluminum is substituted for steel with the same support and loading (unchanged F, A, w, and L) and the deflection δ is not to be increased, what percentage weight saving is possible? (For Al, $\rho = 2.7$ gms/cm³, $E = 10 \times 10^6$ psi, and $\nu = 0.3$, while for steel, $\rho = 7.9$ gms/cm³, $E = 30 \times 10^6$ psi, and $\nu = 0.3$.)

(b) Would the weight saving be greater, less, or the same if both sheets were curved (e.g., corrugated) rather than flat?

12-4. In some design applications, the minimum sheet thickness is controlled by the ability to absorb energy elastically without denting (plastic deformation). In such a case, what percentage weight saving is possible by substituting an aluminum alloy ($Y = 25 \times 10^3$ psi) for a low-carbon steel ($Y = 35 \times 10^3$ psi)? Other properties are given in Prob. 12-3.

12-5. It is necessary to bend an aluminum alloy sheet (0.040 in. thick) to a final radius of curvature of 3 in. The plastic portion of the stress-strain curve may be approximated by $\sigma = (25 \times 10^3) + (25 \times 10^3)\epsilon$, and $E' = 11 \times 10^6$ psi. Accounting for springback, what radius of curvature must be designed into the tools if loading is:

(a) Pure bending?

(b) Tensile enough so the mid-plane of the sheet is stretched 2% in tension?

12-6. What fraction of the cross section remains elastic in Prob. 12-5a?

12-7. It has been suggested that the residual hoop stress in a tube can be determined by slitting a short length of the tube parallel to its axis, measuring the diameter, d, after slitting, and comparing it with the initial diameter, d_o. A distribution of residual stresses must be assumed. Two simple stress distributions are suggested in the sketch. Make analyses for the surface residual stress, σ', using both assumed stress distributions, and compare the predictions for σ' for a copper tube with $d_o = 1.000$ in., $d = 1.005$ in., wall thickness $t = 0.020$ in., $E = 16 \times 10^6$ psi, $\nu = 0.30$.

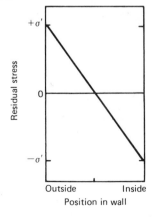

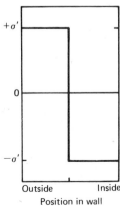

12-8. A round bar (radius R, length L) is plastically deformed in torsion until yielding has occurred throughout the cross section. Assume that the material is ideally plastic (no work hardening), the shear yield stress is k, and the shear modulus is G. When the bar is unloaded it untwists by an amount, $\Delta\theta$ (θ is the angle of twist—i.e., for one revolution $\theta = 2\pi$).

    (a) Derive an expression for the level of residual stress, τ', as a function of radial position r, R, G, and k.

    (b) Find the relative springback, $\Delta\theta/L$ upon unloading in terms of G, k, and R.

13

Plastic anisotropy

13-1 CRYSTALLOGRAPHIC BASIS

With metals, the most important cause of anisotropic plastic properties is the preferred orientation of grains (i.e., a statistical tendency for certain crystallographic orientations)†. Preferred orientations, or crystallographic textures, develop in wrought metals by rotation of the lattice in grains during deformation by slip or twinning. Recrystallization during annealing changes the crystallographic textures and the form of anisotropy, but generally will not produce isotropy. (Repeated heating and cooling through a phase change, such as the $\alpha \longrightarrow \gamma \longrightarrow \alpha$ transformation in steel, is a possible exception.) Details of how crystallographic textures develop and how they impart anisotropy to plastic properties will not be treated here. However, one extreme example, that of textured titanium sheet, will be considered in a qualitative way because of the insight it brings to the problem (see Fig. 13-1).

Alpha-titanium alloys have an hcp crystal structure. Deformation occurs by slip in close-packed $\langle 1\bar{2}10 \rangle$ directions which are perpendicular to the hexagonal or c-axis and parallel to the edges of the hexagonal cell. Though both the (0001) basal plane and the $\{10\bar{1}0\}$ prism planes may act as slip planes, slip on these systems does not produce any strain component parallel to the c-axis. Even if slip occurs on $\{10\bar{1}1\}$ planes which are canted to the c-axis, no c-axis strain will result because the $\langle 1\bar{2}10 \rangle$ direction of slip is normal to the c-axis.

Strong crystallographic textures often develop in rolled and recrystallized

---

†In contrast, anisotropy of fracture behavior is largely governed by mechanical alignment of inclusions, voids, and interfaces.

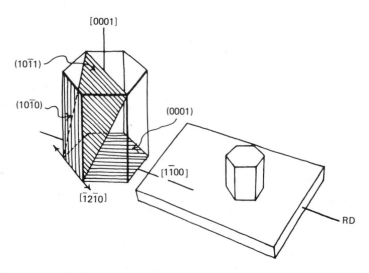

Figure 13-1 Idealized texture of α-titanium with the (0001) basal plane parallel with the plane of the sheet. Although there are three families of potential slip planes, (0001), $\{10\bar{1}1\}$, and $\{10\bar{1}0\}$, slip is restricted to the $\langle 1\bar{2}10 \rangle$ family of directions, all of which lie in the plane of the sheet, so slip causes no thickening or thinning. From W. F. Hosford and W. A. Backofen, in *Fundamentals of Deformation Processing* (Syracuse, N.Y.: Syracuse Univ. Press, 1964), pp. 259–98.

sheets of α-titanium alloys. The texture is most easily described in terms of an ideal orientation in which the basal plane is parallel to the plane of the sheet and the c-axis is aligned with the sheet normal. Of course the orientations of some grains deviate from this ideal orientation†.

Consider the deformation behavior of strip tensile specimens cut from such a rolled sheet. Under uniaxial tension, yielding can occur only by slip on the $\{10\bar{1}0\}$ prism planes. Tensile yield strengths of specimens cut at various angles, θ, to the rolling direction would not vary significantly, and an experimenter might erroneously conclude that the material is isotropic. The error would be apparent if the lateral strains accompanying the tensile extension were also measured. If the orientation is ideal, little or no thickness change will occur, all of the elongation being accompanied by contraction of the width of the tensile specimens. Here a useful parameter is the ratio, R, of plastic strains,

$$R = \frac{\epsilon_w}{\epsilon_t} = \frac{\epsilon_y}{\epsilon_z} \qquad (13\text{-}1)\ddagger$$

as shown in Fig. 13-2 where w and t refer to the width and thickness directions

---

†For α-titanium, the c-axis is rocked $\pm 40°$ from the sheet normal toward the transverse direction.

‡Often the strain ratio is denoted by r. The notation R is used to avoid confusion with radius, which is denoted by r in many sections of this text.

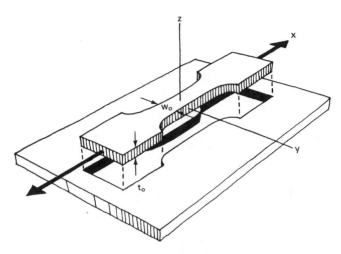

Figure 13-2 Strip tensile specimen cut from a sheet. The R-value is defined as the ratio of the width-to-thickness strains, $\epsilon_w/\epsilon_t = \epsilon_y/\epsilon_z$. From W. F. Hosford and W. A. Backofen, *ibid*.

of the tensile specimen. Thus, $\epsilon_w = \ln(w/w_o)$ and $\epsilon_t = \ln(t/t_o)$. For an isotropic material $R = 1$, while for the ideally oriented titanium, $\epsilon_t = 0$, so $R = \infty$. In commercial titanium sheets, R-values of 3 to 7 are typical, and such high R-values should suggest to the investigator that the material has a high thinning resistance. In fact, a much higher yield stress would be observed if a through-thickness compression test or a balanced biaxial tension ($\sigma_x = \sigma_y$) test† could be made, both of which require thinning strains, because no slip systems are oriented to produce such thinning. Thus, the R-values measured in a strip tension test do indicate in a qualitative way the relative extent of *normal* anisotropy. The yield loci for materials characterized by $R < 1$, $R = 1$, and $R > 1$ are sketched in Fig. 13-3.

13-2 MEASUREMENT OF R

Although the R-value is defined as the ratio of width-to-thickness strains, the thickness strain, ϵ_t, can not be accurately measured in a thin sheet. Therefore the thickness strain is usually found from measurements of length and width strains using volume constancy, $\epsilon_t = -(\epsilon_l + \epsilon_w)$. For accurate measurements, the reduced section should be long relative to its width and the gauge section used for measuring ϵ_l and ϵ_w should be well removed from the shoulders.

 With a few exceptions, the R-value does not vary appreciably with strain. Therefore, the ratio of incremental strains $d\epsilon_w/d\epsilon_t = R$. This constancy of R

---

†The bulge test does impose biaxial tension, $\sigma_x = \sigma_y$, but it is difficult to determine a yield strength at low strains.

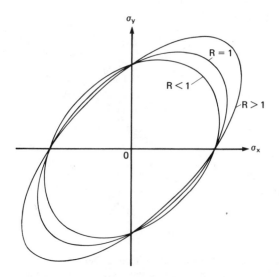

Figure 13-3 Schematic yield loci for textured materials with rotational symmetry. A high R-value implies resistance to thinning and hence a high strength under biaxial tension, while a low R-value indicates easy thinning and therefore a low biaxial strength.

is important when R is used to evaluate the constants in anisotropic yield criteria as in the following sections.

For steels, the R-value and the elastic modulus, E, usually vary similarly with crystallographic texture. Although this correlation is neither fundamental nor exact, it forms the basis for a small commercial instrument that measures E in thin strips by sonic velocity and is calibrated to give readings of the R-value. The R-values usually vary with the test direction, θ, and it is common to characterize a material by an average R-value, $\bar{R}$, where

$$\bar{R} = \frac{R_o + 2R_{45} + R_{90}}{4} \tag{13-2}$$

the subscripts referring to the angle, θ.

13-3 HILL'S ANISOTROPIC PLASTICITY THEORY

Hill[†] has formulated a quantitative treatment of plastic anisotropy without regard to its crystallographic origin. He assumed a homogeneous material characterized by three orthogonal axes of anisotropy, x, y, and z, about which

†R. Hill, *Proc. Roy. Soc. London*, 193A (1948) p. 281 and *Mathematical Theory of Plasticity*, Chapt. XII (London: Oxford Univ. Press, 1950). See also, W. F. Hosford and W. A. Backofen, "Strength and Plasticity of Textured Metals," *Fundamentals of Deformation Processing* (Syracuse, N.Y.: Syracuse Univ. Press, 1964), p. 259.

the properties have two-fold symmetry. (Equivalently, the x-y, y-z and z-x planes are planes of mirror symmetry.) In a rolled sheet, it is conventional to take x, y, and z to be the rolling, transverse, and through-thickness directions respectively. The theory also assumes that in any given direction the tensile and compressive yield strengths are equal.

The proposed anisotropic yield criterion has the form

$$2f(\sigma_{ij}) = F(\sigma_y - \sigma_z)^2 + G(\sigma_z - \sigma_x)^2 + H(\sigma_x - \sigma_y)^2$$
$$+ 2L\tau_{yz}^2 + 2M\tau_{zx}^2 + 2N\tau_{xy}^2 = 1 \qquad (13\text{-}3)$$

where F, G, H, L, M, and N are constants which characterize the anisotropy. Note that if $F = G = H$ and $L = M = N = 3F$, this reduces to the von Mises criterion. The constants F, G, and H can be evaluated from simple tension tests.

Consider such a test in the x direction and let X be the tensile yield stress. At yielding, $\sigma_x = X$, $\sigma_y = \sigma_z = \tau_{ij} = 0$, so Eq. (13-3) becomes $(G + H)X^2 = 1$ or $X^2 = 1/(G + H)$. Similarly, if Y and Z are the tensile yield stresses in the y and z directions,

$$X^2 = \frac{1}{G + H}$$

$$Y^2 = \frac{1}{H + F}$$

$$Z^2 = \frac{1}{F + G} \qquad (13\text{-}4)$$

Solving simultaneously,

$$2F = \frac{1}{Y^2} + \frac{1}{Z^2} - \frac{1}{X^2}$$

$$2G = \frac{1}{Z^2} + \frac{1}{X^2} - \frac{1}{Y^2}$$

$$2H = \frac{1}{X^2} + \frac{1}{Y^2} - \frac{1}{Z^2} \qquad (13\text{-}5)$$

Unfortunately, for sheets it is not convenient to measure Z directly. The constants L, M, and N can be evaluated from shear tests.

The flow rules may be developed using Eq. (2-30),

$$d\epsilon_{ij} = d\lambda \frac{\partial f(\sigma_{ij})}{\partial \sigma_{ij}} \qquad (13\text{-}6)$$

where $f(\sigma_{ij})$ is the yield function. (This applies to anisotropic as well as isotropic materials.) Differentiation of Eq. (13-3) results in the flow rules:

$$d\epsilon_x = d\lambda[H(\sigma_x - \sigma_y) + G(\sigma_x - \sigma_z)], \quad d\epsilon_{yz} = d\epsilon_{zy} = d\lambda L\tau_{yz}$$
$$d\epsilon_y = d\lambda[F(\sigma_y - \sigma_z) + H(\sigma_y - \sigma_x)], \quad d\epsilon_{zx} = d\epsilon_{xz} = d\lambda M\tau_{zx} \qquad (13\text{-}7)$$
$$d\epsilon_z = d\lambda[F(\sigma_z - \sigma_y) + G(\sigma_z - \sigma_x)], \quad d\epsilon_{xy} = d\epsilon_{yx} = d\lambda N\tau_{xy}$$

In deriving the flow rules for the shear strains, $d\epsilon_{yz}$, $d\epsilon_{zx}$, and $d\epsilon_{xy}$, the yield criterion, Eq. (13-3), must be rewritten so that the shear stress terms appear as $L(\tau_{yz}^2 + \tau_{zy}^2) + M(\tau_{zx}^2 + \tau_{xz}^2) + N(\tau_{xy}^2 + \tau_{yx}^2)$; otherwise partial differentiation would lead to the absurd result that $d\epsilon_{yz} = 2 \cdot d\lambda L \tau_{yz}$ and $d\epsilon_{zy} = 0$, etc. Note that for Eqs. (13-7), $d\epsilon_x + d\epsilon_y + d\epsilon_z = 0$, indicating constant volume. Consider an x-direction tension test again. Substituting $\sigma_x = X, \sigma_y = \sigma_z = 0$ into Eq. (13-7) gives the resulting strains, $d\epsilon_x = d\lambda(H + G)X$, $d\epsilon_y = -d\lambda(H)X$, $d\epsilon_z = -d\lambda(G)X$. Since the strain ratio for the x-direction tension test is defined as $R = R_0 = (d\epsilon_y/d\epsilon_z)$,

$$R = \frac{H}{G} \tag{13-8}$$

Similarly, defining $P = R_{90}$ as the strain ratio in a y-direction tension test, $P = d\epsilon_x/d\epsilon_z$. With $\sigma_y = Y$ and $\sigma_x = \sigma_z = 0$, Eqs. (13-7) result in

$$P = \frac{H}{F} \tag{13-9}$$

Equations (13-8) and (13-9) allow one to predict the value of the z-direction yield stress, Z, by conducting x- and y-direction tension tests and measuring R and P as well as X and Y. According to Eq. (13-4), $Z^2 = 1/(F + G)$ and $X^2 = 1/(G + H)$, so

$$\frac{Z^2}{X^2} = \frac{(G + H)}{(F + G)} = \frac{(1/R) + 1}{1/R + 1/P}$$

or

$$Z = X\sqrt{P(1 + R)/(P + R)}$$

Similarly,

$$Z = Y\sqrt{R(1 + P)/(P + R)} \tag{13-10}$$

13-4 SPECIAL CASE IN WHICH x, y, z
ARE PRINCIPAL STRESS AXES

For loading conditions in which x, y, and z are principal stress axes, Hill's yield criterion can be restated in terms of R and P. Letting $\tau_{yz} = \tau_{zx} = \tau_{xy}$ be zero in Eq. (13-3), substituting $1 = (G + H)X^2$ from Eq. (13-4), and dividing by G,

$$\left(\frac{F}{G}\right)(\sigma_y - \sigma_z)^2 + \left(\frac{G}{G}\right)(\sigma_z - \sigma_x)^2 + \left(\frac{H}{G}\right)(\sigma_x - \sigma_y)^2 = \left[\left(\frac{G}{G}\right) + \left(\frac{H}{G}\right)\right]X^2$$

Now substituting $R = H/G$ and $R/P = F/G$ and multiplying by P,

$$R(\sigma_y - \sigma_z)^2 + P(\sigma_z - \sigma_x)^2 + RP(\sigma_x - \sigma_y)^2 = P(R + 1)X^2 \tag{13-11}$$

Similarly, the flow rules, Eqs. (13-7), reduce to

$$d\epsilon_x : d\epsilon_y : d\epsilon_z = R(\sigma_x - \sigma_y) + (\sigma_x - \sigma_z) : \left(\frac{R}{P}\right)(\sigma_y - \sigma_z) + R(\sigma_y - \sigma_x):$$

$$\frac{R}{P}(\sigma_z - \sigma_y) + (\sigma_z - \sigma_x) \tag{13-12}$$

13-5 SPECIAL CASE OF ROTATIONAL SYMMETRY ABOUT z

If the material has rotational symmetry about the z-axis (planar isotropy), $F = G, L = M$, and $R = P$. For the loading condition where z is a principal axis (i.e., $\tau_{yz} = \tau_{zx} = 0$) and with rotational symmetry, x and y may be chosen to coincide with the other two principal stress axes so that $\tau_{xy} = 0$. In this case, substitution of $P = R$ in Eqs. (13-11) and (13-12) results in

$$(\sigma_y - \sigma_z)^2 + (\sigma_z - \sigma_x)^2 + R(\sigma_x - \sigma_y)^2 = (R + 1)X^2 \tag{13-13}$$

and

$$d\epsilon_x : d\epsilon_y : d\epsilon_z = (R + 1)\sigma_x - R\sigma_y - \sigma_z : (R + 1)\sigma_y - R\sigma_x - \sigma_z : 2\sigma_z - \sigma_x - \sigma_y$$

$$\tag{13-14}$$

A further simplification results for the case of plane-stress ($\sigma_z = 0$) loading. The yield criterion is then

$$\sigma_x^2 + \sigma_y^2 - \frac{2R}{(R + 1)}\sigma_x\sigma_y = X^2 \tag{13-15}$$

or

$$\frac{\sigma_x^2}{X^2} = \left(1 + \alpha^2 - \frac{2\alpha R}{R + 1}\right)^{-1} \tag{13-16}$$

where α is the stress ratio, $\alpha = \sigma_y/\sigma_x$.

Equation (13-15) plots as an ellipse whose major and minor axes depend upon R, as shown in Fig. 13-4. With $R = 1$, the standard von Mises ellipse results. It is obvious that higher values of R lead to increased resistance to yielding under biaxial tension. This forms the basis for a considerable texture hardening effect. For example, if $\alpha = 1$ (i.e., balanced biaxial tension, $\sigma_x = \sigma_y$) and $R = 5, \sigma_x/X = \sqrt{3} = 1.732$, or a 73 % increase over what would be expected from an isotropic material ($R = 1$).

13-6 GENERALIZATION OF HILL'S CRITERION

It is possible to calculate the shape of yield loci for crystallographically textured fcc and bcc metals. Such calculations† for a wide range of textures indicate that the Hill yield criterion tends to overestimate the effect of the R-value on

---

†W. F. Hosford, "On Yield Loci of Anisotropic Cubic Metals," *7th North American Metalworking Research Conference Proceedings*, SME, Dearborn, Mich. (1979) p. 191; and R. Logan and W. F. Hosford, *Int. Jour. Mech. Sci.*, 22 (1980) pp. 419–30.

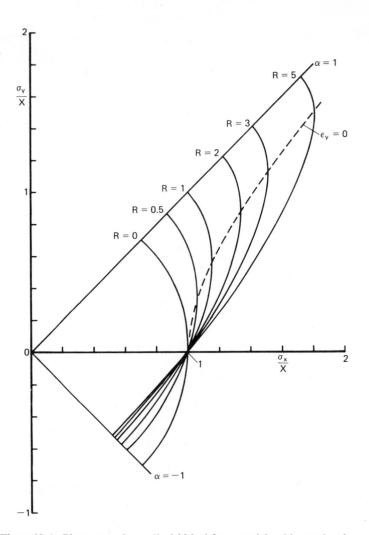

Figure 13-4 Plane-stress ($\sigma_z = 0$) yield loci for materials with rotational symmetry about z, according to the Hill criterion, Eq. (13-13). Dashed line is the locus of conditions for plane strain, $\epsilon_y = 0$.

the shape of the yield loci. The trends of these calculations are better represented by a generalization of Hill's criterion of the form

$$F|\sigma_y - \sigma_z|^a + G|\sigma_z - \sigma_x|^a + H|\sigma_x - \sigma_y|^a = 1 \qquad (13\text{-}17)$$

where the exponent a is much larger than the 2 in Hill's criterion. For planar isotropy and plane-stress loading this simplifies to

$$|\sigma_x|^a + |\sigma_y|^a + R|\sigma_x - \sigma_y|^a = (R+1)Y^a \qquad (13\text{-}18)$$

The calculations suggest that $a \simeq 6$ for bcc metals and $a \simeq 8$ to 10 for fcc metals. Figure 13-5 shows the yield loci predicted by Eq. (13-18) for several

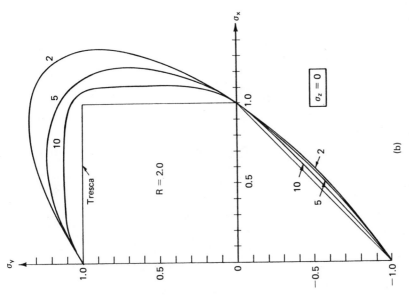

(b)

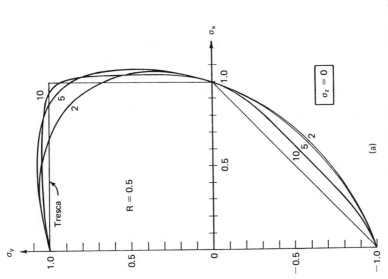

(a)

Figure 13-5 Plane-stress ($\sigma_z = 0$) yield loci for materials with rotational symmetry about z for the generalized yield criterion, Eq. (13-18). Exponents higher than $a = 2$ (Hill criterion) predict less effect of R upon the shape of the locus. From W. F. Hosford, in *7th North Amer. Metalworking Res. Conf. Proc.*, SME, Dearborn, Mich. (1979), pp. 191–96. See also, R. W. Logan and W. F. Hosford, *Int. J. Mech. Sci.*, 22, no. 7 (1980), pp. 419–30.

values of a and R. Note that with increasing values of a, the loci approach the Tresca locus. The corresponding flow rule for $\sigma_x > \sigma_y$ is

$$\frac{d\epsilon_y}{d\epsilon_x} = \frac{|\sigma_y|^{a-1} - R(\sigma_x - \sigma_y)^{a-1}}{|\sigma_x|^{a-1} + R(\sigma_x - \sigma_y)^{a-1}} \qquad (13\text{-}19)$$

PROBLEMS

13-1. A tensile specimen is cut at an angle θ to the x-axis (rolling direction). Show that the Hill criterion and flow rules give the R_θ value as

$$R_\theta = \frac{H + (2N - F - G - 4H)\sin^2\theta\cos^2\theta}{F\sin^2\theta + G\cos^2\theta}$$

13-2. (a) Using the results of Prob. 13-1, derive an expression for N/G in terms of R, $P(= R_{90})$ and R_{45}.

(b) Strip tension tests on a rolled sheet in rolling (x) and transverse (y) directions produced the following results:

rolling direction, Y.S. = 71,500 psi, $R_o = 4.0$

transverse direction, Y.S. = 65,300 psi, $R_{90} = 2.0$

In a strip tension test at 45°, a strain ratio, $R_{45} = 2.5$ was recorded. Calculate, with the Hill theory, the yield stresses at $\theta = 22\frac{1}{2}°$, $\theta = 45°$, $\theta = 67\frac{1}{2}°$ and plot the Y.S. versus θ.

13-3. Is it possible for a sheet material to have a minimum or maximum of the Y.S. versus θ curve between $\theta = 0$ and $\theta = 90$? If so, what relationship must exist between F, G, H, L, M, and N?

13-4. A thin-walled tube is made from a sheet of metal by bending the sheet into a cylinder and welding it; the prior rolling direction becomes the axial direction of the tube and the prior transverse direction becomes the hoop direction. The tube has a diameter of 5.00 in. and a wall thickness of 0.025 in. The strain ratios in the sheet are $R_o = 2.5$, $R_{45} = 1.8$, and $R_{90} = P = 0.8$, and the yield strength in the rolling direction is 100,000 psi. Neglect elastic effects.

(a) If the tube is capped and loaded under internal pressure, at what pressure will the walls yield?

(b) Will the length of the tube increase, decrease, or remain constant? Calculate the ratio of $\epsilon_{axial}/\epsilon_{hoop}$.

(c) If the tube is extended axially by a tensile force, will the volume *inside* the tube increase, decrease, or remain constant?

(d) If the capped tube is filled with an incompressible fluid (water) and pulled in tension, what will be the tensile stress at yielding?

13-5. Consider an anisotropic sheet material for which the properties in all directions in the plane of the sheet are the same, loaded in plane stress (σ_x, σ_y) with $\sigma_z = 0$.

(a) Express the strain ratio $\rho = \epsilon_y/\epsilon_x$ as a function of the stress ratio $\alpha = \sigma_y/\sigma_x$ and R.

(b) Write an expression for $\bar{\sigma}$ in terms of α, R, and σ_x. Define $\bar{\sigma}$ and $\bar{\epsilon}$ such that $\bar{\sigma}$ and $\bar{\epsilon}$ reduce to σ_x and ϵ_x in an x-direction tension test.

(c) Write an expression for $d\bar{\epsilon}$ in terms of α, R, and $d\epsilon_x$. Remember that $\bar{\sigma}d\bar{\epsilon}$
$= \sigma_x d\epsilon_x + \sigma_y d\epsilon_y + \sigma_z d\epsilon_z$.

13-6. Using the Hill criterion for a metal sheet without rotational symmetry,
(a) Derive an expression for $\alpha = \sigma_y/\sigma_x$ for plane-strain ($\epsilon_y = 0$) and plane-stress ($\sigma_z = 0$) loading.
(b) Find the stress, σ_x, for this plane-strain yielding in terms of X, R, and P.
(c) Evaluate α and σ_x, if $R = 2.0$, $P = 1.5$, and $X = 30,000$ psi.
(d) Predict the y-direction yield strength, Y, for this sheet.

13-7. Tensile specimens cut from a sheet in the x and y directions have strain ratios of $R = 1.6$ and $P = 2.0$, and the yield strength in the x direction is 50,000 psi.
(a) Calculate the yield strength for the y-direction test.
(b) The material is deformed in a plane-strain tension test with $\sigma_z = 0$, $\epsilon_y = 0$. Calculate σ_x at yielding.
(c) The material is deformed such that $\epsilon_z = 0$, $\sigma_z = 0$. Calculate σ_x and σ_y at yielding.
(d) The material is deformed such that $\epsilon_x = \epsilon_y$ and $\sigma_z = 0$. Calculate σ_x and σ_y at yielding.

13-8. (a) For sheet materials with rotational symmetry about z, and $\bar{R} = 0.5$, calculate the stress ratio for plane strain, $\epsilon_y = 0$, according to Hill, and according to the generalized anisotropic yield criterion, Eq. (13-18) with $a = 6$.
(b) Repeat for $\bar{R} = 2.0$.

14

Cupping, redrawing, and ironing

14-1 INTRODUCTION

Sheet-metal forming processes, broadly classified as deep drawing or stamping operations, represent a wide spectrum of flow conditions. The material properties that control formability vary with the specific sheet-forming operation. At one end of the spectrum is the forming of flat-bottom cylindrical cups by radial drawing (i.e., cupping). In this case one of the principal strains in the plane of the sheet is positive and the other is negative, the change of thickness being small. At the other end of the spectrum are operations involving biaxial stretching of the sheet, where two of the principal strains are tensile, and thinning is required. Many operations fall between these extremes. In forming many parts, stretching may predominate in one region while drawing prevails in another.

Sheet forming differs from most bulk-forming processes in several respects. In sheet, tension predominates, whereas bulk-forming operations are mainly compressive. Furthermore, one or both surfaces of deforming regions are often free (i.e., not supported by tools).

For both types of processes, formability depends upon lubrication and tooling, as well as on material properties. The major difference is that failure in sheet forming is usually by plastic instability in tension rather than fracture. Even though a sheet may finally break, the useful forming limit is reached when a localized neck forms, rather than by the fracture event itself. Admittedly, it is possible to find instances in sheet forming where fracture precedes any necking, but these instances are rare. Neither of the two common measures of tensile ductility, namely reduction in area or total elongation, correlate well with press performance, though the latter may be a useful index in some operations, since it does depend upon uniform elongation as well as the rate at which a neck localizes.

14-2 CUP DRAWING—MATERIAL EFFECTS

Consider the deep drawing of cylindrical flat-bottom cups (cupping or Swift cup testing)† This is a relatively simple process at one extreme end of the sheet-forming spectrum; it is used commercially to produce such items as steel pressure and vacuum vessels, zinc dry cells, brass flashlights and cartridge cases, and aluminum and steel beer cans. Figure 14-1 illustrates cupping and defines

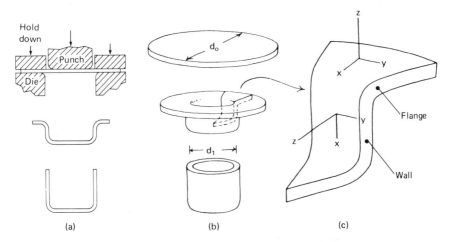

Figure 14-1 Schematic illustration of cupping and coordinate axes.

the coordinate system x, y, and z. There are clearly two important regions: the flange where most of the deformation occurs, and the wall which must support a sufficient force to cause the deformation in the flange. If the blank diameter is too large, the force that must be transmitted by the wall will be excessive, thereby causing it to yield and fail. The formability may be expressed as a limiting drawing ratio (LDR) which is the largest ratio of blank-to-cup diameters, (d_o/d_1), that may be drawn successfully. In some respects cupping is similar to wire drawing. Indirect compression is generated in the flange (analogous to the wire die) by tension in the drawn material. Each wedge-shaped element of the initial blank must be sufficiently compressed in the circumferential direction to permit it to flow over the die lip. As in wire drawing, if the reduction is too large, the tensile stresses in the drawn section will exceed the *tensile* strength and cause failure. Figure 14-2 illustrates such a failure. The

†S. Y. Chung and H. W. Swift, *Proc. Inst. Mech. Eng.*, 165 (1951), p. 199.

principal difference is that wire drawing is a steady-state process while cupping is not. The function of the hold-down force is simply to prevent wrinkling in the flange. (Recall the wrinkling in aluminum foil pie dishes, which are made by a similar process with no hold-down.) Wrinkles in the flange of a partially drawn cup are shown in Fig. 14-3.

Figure 14-2 Drawing failures by neck-ing at bottom of cup wall. With very low friction, the failure site tends to move onto the punch radius as shown at the right. From D. J. Meuleman, Ph.D. thesis, Univ. of Michigan (1980).

Figure 14-3 Wrinkling in a partially drawn cup due to insufficient hold-down force. From D. J. Meuleman, *ibid*.

The following analysis, based largely on the work of Whiteley†, shows that the LDR depends primarily upon the average strain ratio, $\bar{R}$. The coordinate system is illustrated in Fig. 14-4.

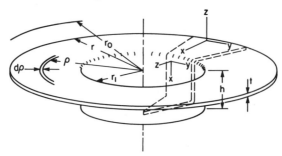

Figure 14-4 Schematic illustration of a partially drawn cup showing coordinate system. From W. F. Hosford, in *Formability; Analysis, Modeling and Experimentation*, AIME (1978), pp. 78–95.

A number of simplifying assumptions are invoked:

1. All of the energy expended is used to deform the material in the flange. Work against friction and work to bend and unbend the sheet as it flows

†R. Whiteley, *Trans. ASM*, 52 (1960), p. 154. See also, W. F. Hosford, "Effect of Anisotropy and Work Hardening on Cup Drawing, Redrawing and Ironing," *Formability; Analysis, Modeling and Experimentation*, Metallurgical Society of AIME, N.Y. (1978), p. 78.

over the die lip are neglected, in the initial treatment, but will be accounted for later by an efficiency factor.

2. The material does not work harden ($n = 0$). It will be shown later that the n-value has only a minor effect on the LDR.

3. The flow in the flange is characterized by plane strain, $\epsilon_z = 0$, so the wall thickness of the cup will be the same as that of the starting blank. Although some wall thinning occurs near the base of the cup and thickening near the top, these changes are not large and this assumption greatly simplifies the analysis.

4. The material properties are rotationally symmetric, so that there is "planar isotropy" and "normal anisotropy." Later, in applying the theory to real materials, it will be assumed that angular variations of R can be handled by using the average-strain ratio, $\bar{R} = (R_o + 2R_{45} + R_{90})/4$.

5. Yielding is described by Hill's anisotropic plasticity theory.

Consider first the deformation of the flange (see Fig. 14-4). With the assumption that $d\epsilon_z = 0$, the total surface area remains unchanged, so the area *inside* any element initially at a distance ρ_o from the center is constant. Therefore,

$$\pi \rho^2 + 2\pi r_1 h = \pi \rho_o^2$$

so

$$2\pi \rho \, d\rho + 2\pi r_1 \, dh = 0$$

or

$$d\rho = \frac{-r_1 \, dh}{\rho} \tag{14-1}$$

The circumference of the element is proportional to ρ, so $d\epsilon_y = d\rho/\rho$.
Since $d\epsilon_z = 0$,

$$d\epsilon_x = -d\epsilon_y = \frac{-d\rho}{\rho} = \frac{r_1 dh}{\rho^2} \tag{14-2}$$

where r_1 is the punch radius and dh is the incremental distance moved by the punch. The incremental work done on this element is equal to the volume of the element, $(2\pi t \rho \, d\rho)$, times the incremental work per volume, $(\sigma_x \, d\epsilon_x + \sigma_y \, d\epsilon_y + \sigma_z \, d\epsilon_z)$. Since $d\epsilon_z = 0$, $d\epsilon_x = -d\epsilon_y$, the work per volume is $(\sigma_x - \sigma_y)d\epsilon_x$, so the work on this element is $dW = 2\pi t \rho \, d\rho(\sigma_x - \sigma_y)r_1 dh/\rho^2$.

Although the relative values of σ_x and σ_y vary with element position, the term $(\sigma_x - \sigma_y)$ should be constant and is designated by σ_f (flow strength of the flange under the constraint $d\epsilon_z = 0$). With $d\epsilon_z = 0$ and $\sigma_z = 0$, $\sigma_y = -\sigma_x$, so $\sigma_f = 2\sigma_x$. The total work on all such elements per increment of punch travel is

$$\frac{dW}{dh} = \int_{r_1}^{r} \frac{2\pi r_1 t \sigma_f \, d\rho}{\rho} = 2\pi r_1 t \sigma_f \ln\left(\frac{r}{r_1}\right) \tag{14-3}$$

The drawing force, F_d, which must equal dW/dh, will have its largest value at the beginning of the draw when $r = r_o$, so

$$F_{d(\text{max})} = 2\pi r_1 t\sigma_f \ln\left(\frac{r_o}{r_1}\right) = 2\pi r_1 t\sigma_f \ln\left(\frac{d_o}{d_1}\right) \tag{14-4}$$

where d_o and d_1 are the blank and punch diameters respectively.

In the cup wall, which must carry the force $F_{d(\text{max})}$, the axial stress, $\sigma_x = F_{d(\text{max})}/2\pi r_1 t$, is

$$\sigma_x = \frac{F_{d(\text{max})}}{2\pi r_1 t} = \sigma_f \ln\left(\frac{d_o}{d_1}\right) \tag{14-5}$$

Since it was assumed that $n = 0$, the walls start to neck as soon as they yield. Therefore, the drawing limit will be reached when this stress reaches the flow strength of the wall, σ_w, i.e., when

$$\sigma_w = \sigma_f \ln\left(\frac{d_o}{d_1}\right) \tag{14-6}$$

or
$$\ln(\text{LDR}) = \ln\left(\frac{d_o}{d_1}\right) = \frac{\sigma_w}{\sigma_f} \tag{14-7}$$

The flow strength, σ_w, is characterized by plane strain, $\epsilon_y = 0$, because the wall circumference is constrained by the punch from shrinking. Thus, the LDR is governed by the ratio, β, of the two plane-strain flow strengths,

$$\beta = \frac{\sigma_{w(\epsilon_y=0)}}{\sigma_{f(\epsilon_z=0)}} \tag{14-8}$$

For an isotropic material, $\beta = 1$, ($\sigma_f = \sigma_w$), so Eq. 14-7 predicts LDR $= e = 2.72$. In practice, the drawing limit is nearer 2.1 to 2.2. The reason that the theoretical value is too high is that friction and bending have been neglected. However, the development can be modified by a deformation efficiency, η, to account for frictional and bending work.† The total work per increment of punch travel, dW/dh, would then be $1/\eta$ times as large as that predicted by Eq. 14-3, and Eq. 14-7 would become

$$\ln(\text{LDR}) = \eta\beta \tag{14-9}$$

The efficiency, η, will vary with lubrication, hold-down pressure, sheet thickness, and die radius, but a typical value can be estimated from the fact that the LDR values for relatively isotropic materials are in the range of 2.1 to 2.2; this gives $\eta = 0.74$ to 0.79.

At this point Whiteley used Hill's anisotropic plasticity theory to relate

---

†In Whiteley's paper, the symbol η is used to designate the ratio of frictional and bending work to ideal work rather than an efficiency. In his notation, $1/(1 + \eta)$ is equivalent to an efficiency factor. The notation has been altered here to make it consistent with that used in other areas of metal forming. It should be noted that here η is the ratio of $dW_{\text{flange}}/dW_{\text{total}}$ at the point of F_{max}.

β to the more easily measured R. In the flange, where $d\epsilon_z = 0$, the flow rules, Eq. (13-14), predict that $\sigma_z = (\sigma_x + \sigma_y)/2$. Substituting this into the yield criterion, Eq. (13-13), results in

$$\frac{(\sigma_y - \sigma_x)^2}{4} + \frac{(\sigma_y - \sigma_x)^2}{4} + R(\sigma_x - \sigma_y)^2 = (1 + R)X^2$$

or $\qquad\qquad \sigma_f = (\sigma_x - \sigma_y)_{(\epsilon_z = 0)} = X\left(\frac{1+R}{R+\frac{1}{2}}\right)^{1/2}$ $\qquad\qquad$ (14-10)

In the wall, where $\epsilon_y = 0$ and $\sigma_z = 0$, the flow rules predict that $\sigma_y = R\sigma_x/(R + 1)$, so the yield criterion reduces to

$$\sigma_w = \sigma_{x(\epsilon_y = 0, \sigma_z = 0)} = X\frac{1+R}{\sqrt{2R+1}} \qquad\qquad (14\text{-}11)$$

Substituting Eqs. (14-10) and (14-11) into the definition of β, Eq. (14-8), the Hill theory predicts that

$$\beta = \sqrt{(R+1)/2} \qquad\qquad (14\text{-}12)$$

so Eq. (14-9) becomes

$$\ln(\text{LDR}) = \eta \sqrt{(R+1)/2} \qquad\qquad (14\text{-}13)$$

Although this development assumed rotational symmetry of properties and therefore a unique strain ratio, R does vary with direction in real sheets. To compare theory and experiment, it has been common to use the average strain ratio, $\bar{R}$, defined by Eq. (13-2), so

$$\ln(\text{LDR}) = \eta \sqrt{(\bar{R}+1)/2} \qquad\qquad (14\text{-}14)$$

Experimental results shown in Fig. 14-5 indicate that the LDR does increase with increasing $\bar{R}$. The dashed lines represent Eq. 14-14 with η adjusted to give a best fit. Clearly, this theory predicts a much greater dependence of the LDR on $\bar{R}$ than that found by experiments.

Probably the largest source of error is the use of the Hill criterion to characterize the anisotropy. Figure 14-6 shows values of R and β calculated for various rotationally symmetric textures for fcc metals.† Although there is much scatter, the R-β trend (dashed line) is better represented by

$$\beta = \left(\frac{2R}{R+1}\right)^{0.27} \qquad\qquad (14\text{-}15)$$

than the Hill criterion (solid line). Substitution of Eq. (14-15) into Eq. (14-9), with the efficiency chosen to give the best fit, leads to the solid lines in Fig. 14-5.

---

†W. F. Hosford and C. Kim, *Met. Trans.*, 7A (1976), pp. 468–72. For rotational symmetry, $R = \bar{R}$.

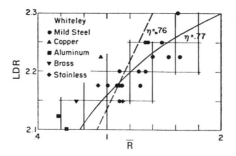

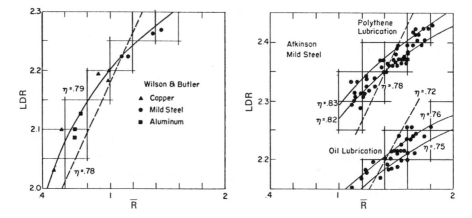

Figure 14-5 Dependence of limiting drawing ratio on $\bar{R}$. Dashed lines are predictions of Eq. (14-14) and solid lines are from substitution of Eq. (14-15) into Eq. (14-9). For both, η was adjusted for the best fit. From W. F. Hosford and C. Kim, *Met. Trans.*, 7A (1976), p. 468; experimental data from: R. L. Whiteley, *ibid*; D. V. Wilson and R. D. Butler, *J. Inst. Metals*, 90 (1963), p. 473; and M. Atkinson, *Sheet Metal Ind.*, 41 (1964), p. 167.

14-3 EFFECTS OF WORK HARDENING

The effects of work hardening can be incorporated into the previous analysis. Instead of describing the mathematical details, only the results will be given here. Figure 14-7 shows the calculated variation of LDR with the strain hardening exponent, n, for several levels of β, not R. It can be seen that the effect of n is rather minor, especially in the normal range of $0.2 \leq n \leq 0.5$.

Work hardening also affects the variation of the punch force during the stroke. Figure 14-8 shows calculations of punch force versus stroke for several levels of n. As n increases, the maximum force occurs later in the stroke. The

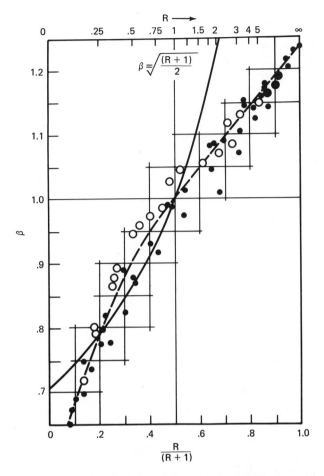

Figure 14-6 The calculated relation of R and β for radially symmetric sheets deforming by $\{111\}\langle110\rangle$ or $\{110\}\langle111\rangle$ slip. The solid points are for textures with a single (hkl) plane parallel to the sheet. Open circles are for textures with varying proportions of $\{100\}$, $\{110\}$, and $\{111\}$ lying nearly parallel to the sheet. The dashed trend through the calculated points is $\beta = [2R/(R+1)]^{0.27}$ while the solid line is from Hill's theory, Eq. (14-12). From W. F. Hosford and C. Kim, *ibid.*

calculations neglect the influence of the curvatures of the die lip and punch which permit some stroke to occur before truly cylindrical walls form. As a consequence, in practice the force initially rises less rapidly with stroke and the maxima occur later than shown here.

To simplify the previous analyses, it was assumed that no thickness change occurs during drawing; this is not really true. The bottom of the cup wall is thinned while the top is thickened. Figure 14-9a shows the calculated wall

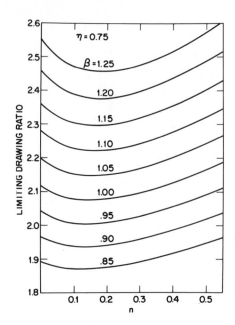

Figure 14-7 Calculated values of LDR for different strain-hardening exponents with $\eta = 0.75$. From W. F. Hosford, *ibid.*

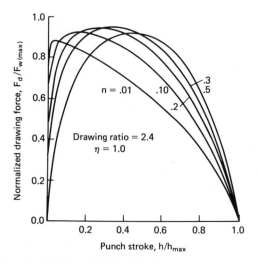

Figure 14-8 Calculated variation of drawing force with stroke, *h*. The drawing force is normalized by the wall strength and the punch stroke by the final cup height. Note that the maximum punch force is reached later as the *n*-value increases. From W. F. Hosford, *ibid.*

thickness changes for an isotropic non-work hardening material, while experimental measurements are given in Fig. 14-9b.

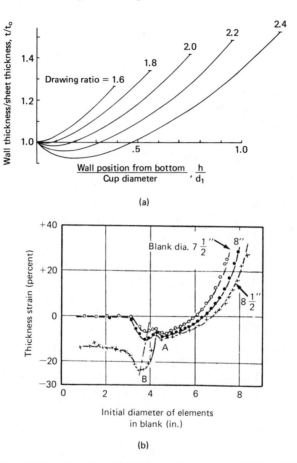

(a)

(b)

Figure 14-9 (a) Variation of wall thickness with position (calculated), (b) variation of wall thickness with position (experimental). From W. Johnson and P. B. Mellor, *Engineering Plasticity* (New York: Van Nostrand Reinhold, 1973), p. 297. See also, S. Y. Chung and H. W. Swift, *Proc. Inst. Mech. Eng.*, 165 (1951), pp. 199–228.

14-4 DEFORMATION EFFICIENCY

Some defense should be made for the simplistic use of an efficiency factor to treat the contributions of friction, dW_f/dh, and bending-unbending, dW_b/dh, to the total drawing force. For analysis of drawability, it is important only that dW_f/dh and dW_b/dh are proportional to the maximum drawing force,

$F_{d(\text{max})}$, at the drawing limit. The bending-unbending at the die lip will be in plane strain, $d\epsilon_y = 0$, so the forces arising from bending should be proportional to $F_{d(\text{max})}$, which also relates to $d\epsilon_y = 0$.

Friction arises in two places, one being on the die lip. There the normal force, N, acting against the die lip is proportional to the drawing force, $N \approx (\pi/2)F_d$, so the frictional force is approximately $\pi\mu F_d/2$ (see Prob. 14-4). Friction is also important in the flange, where the normal force is the hold-down force. If the hold-down force is a constant fraction of the maximum drawing force, $F_{d(\text{max})}$, frictional forces will be in proportion to $F_{d(\text{max})}$. It has been suggested† that the hold-down *pressure* be a constant fraction of the tensile strength, which is a simpler procedure and is almost equivalent. With high R- (or β) values, the maximum drawing stress will be raised above the tensile strength. However, the increase in drawability with high R (or β) will also allow greater flange areas, so the hold-down force (hold-down pressure times area) will also be increased roughly in proportion to the increase of the maximum drawing force. Thus the value of η at the drawing limit should not vary significantly with material properties.

Several investigators‡ have considered another mode of failure for materials of low n-value. Analytical predictions indicate that early in the punch stroke, before any wall is formed, a necking failure would occur in the unsupported region between the punch and die lip. No attempt has been made here to treat this failure mode or its effect on the limiting drawing ratio. However, it should be recognized that such failures would occur so early in the draw that little work would be expended against friction or bending over the die lip, so the mechanical efficiency should be much higher than in later stages after a wall has been formed. Therefore, these analyses, which are based on a 100% efficiency, underestimate the drawing force later in the draw. Thus in practice, the limiting drawing ratio may be reached at blank sizes smaller than those required to produce these early failures.

14-5 CUP DRAWING—EFFECTS OF TOOLING

Up to this point, emphasis has been placed upon material properties on cupping. However, tooling and lubrication also play an important role. The work expended in bending and unbending the sheet as it flows over the die lip increases with the ratio of the sheet thickness to radius of curvature of the die lip, causing

†Chung and Swift, *ibid.*

‡D. C. Chiang and S. Kobayashi, *J. Eng. Indust.*, 88 (1966), p. 443; B. Budiansky and N. M. Wang, *J. Mech. Phys. Solids*, 14 (1966), p. 357; M. G. El-Sebaie and P. B. Mellor, *Int. J. Mech. Sci.*, 15 (1973), p. 485.

higher drawing forces and lower LDRs.† However, if the die radius is too large, wrinkling of the sheet can occur in the unsupported region between the die and the punch.

The punch-nose radius is also important. The failure site is normally at the bottom of the cup wall where the radiused portion meets the cylindrical surface, and the wall is weakest here because it has been work hardened the least. With more generous punch radii, this failure site is moved upward into material that has been strengthened by more prior work hardening.

The effect of friction is twofold. Lubrication of the flange is beneficial since it reduces the work expended to overcome friction. On the other hand, high friction on the cylindrical surface of the punch can increase drawability. With any stretching of the wall, elements on the wall would move upward relative to the punch, causing a shear stress between wall and punch so that the bottom of the wall does not experience the full drawing force. This is important because the lower wall has not been work hardened as much as regions further up the wall. Thus, roughened punches and differential lubrication can be used to increase drawability. It may, however, be difficult to preserve these benefits in production because of punch wear and accidental transfer of lubricant to the punch. A similar transfer of the drawing force to the sides of the punch (thereby protecting the bottom of the wall from the full drawing force) may be aided by drawing into a die cavity under high pressure, which forces the wall against the punch‡. With very low friction there is increased thinning of the cup bottom, and the failure site tends to shift downward into the punch radius, as shown in Fig. 14-2.

Some increase in drawability has been achieved by maintaining the punch at a temperature lower than the flange.§ This temperature differential serves to lower the flow stress of the flange relative to that of the wall.

Ideally, the hold-down pressure should be adjusted to a level just sufficient to prevent wrinkling of the flange, since flange friction increases with hold-down pressure. Swift** recommended a hold-down pressure of $\frac{1}{2}$ to 1% of the yield stress, but the optimum hold-down pressure decreases with increasing thickness-to-diameter ratio, t/d, of the blank.†† Above $t/d = 0.025$, no hold-down is required. For very thin sheets the hold-down pressure necessary to prevent wrinkling is so large that the increased friction lowers the LDR.

†Chung and Swift, *ibid.*

‡W. G. Granzow, "Sheet Metal Forming Using the Aquadraw Process," Paper No. F-1920, Fabricating Machinery Assn., Rockford, Ill. (1975), pp. 1–18.

§W. G. Granzow, "The Effect of Tooling Temperature on Formability of Low Carbon Steel Sheets," Sheet Metal Industries, 55 (1979), p. 561.

**Chung and Swift, *ibid.*

††D. F. Eary and E. A. Reed, *Techniques of Pressworking Sheet Metal, 2nd Ed.* (Englewood Cliffs, N.J.: Prentice-Hall, 1974).

The top edges of drawn cups are not completely flat. Rather, there are high points or *ears* with valleys between (see Fig. 14-10). Four ears are most common but occasionally two, six, or even eight ears may be found. Earing is due to planar anisotropy, and correlates well with the angular variation of R as shown schematically in Fig. 14-11. At angular positions with low R-values, more thickening occurs, so the wall heights are lower; at positions of high R, the walls are thinner and higher. The ear height and position correlate well with

Figure 14-10 Earing behavior of cups made from three different copper sheets. Arrow indicates rolling direction of the sheets. From D. V. Wilson and R. D. Butler, *J. Inst. Met.*, 90 (1961–2), pp. 473–83.

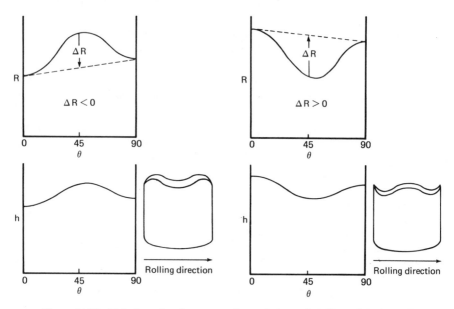

Figure 14-11 Relation of earing to angular variations of R. Here, h is the wall height.

the parameter

$$\Delta R = \frac{R_o + R_{90} - 2R_{45}}{2}$$

This is demonstrated in Fig. 14-12. When $\Delta R > 0$, ears form at $0°$ and $90°$, while if $\Delta R < 0$, ear formation occurs near $\pm 45°$. Earing is undesirable since more metal must be trimmed. Therefore, the full benefit of a high $\bar{R}$ value on LDR may not be realized if ΔR is too large.

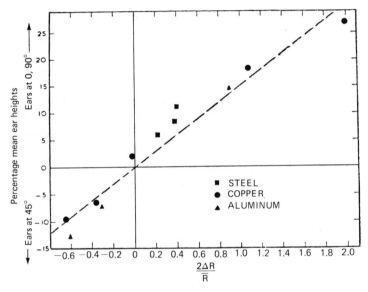

Figure 14-12   Correlation of extent of earing with ΔR. From D. V. Wilson, and R. D. Butler, *ibid*.

14-7 REDRAWING

A single cupping operation will usually not produce a cup deep enough for most applications. However, the diameter can be reduced and the wall height increased by redrawing. In direct redrawing, hold-down is provided by a sleeve, while in reverse redrawing, inversion of the cup presses it against the die, and hold-down is often unnecessary for preventing wrinkling. Reverse redrawing, however, is not readily adaptable to high production rates.

Figure 14-13 illustrates direct and reverse redrawing. Redrawing, like cupping, is characterized by a limiting reduction; if the redraw ratio is too large, the cup wall will fail in tension, yet there are several differences between these two processes. In cupping, the maximum punch force is experienced early in the stroke and then falls off as the amount of material being deformed decreases. On the other hand, redrawing is almost a steady-state process.

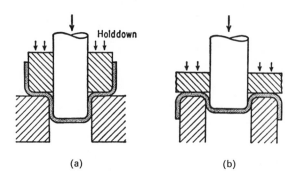

Holddown

(a)   (b)

Figure 14-13   Direct redrawing (a) and reverse redrawing (b). From G. E. Dieter, Jr., *Mechanical Metallurgy, 2nd Ed.* (New York: McGraw-Hill, 1976), p. 692.

If the material were non-work hardening and the wall of the initial cup were of constant thickness, the punch force would be constant during the redrawing stroke (see Fig. 14-14). However, the top of the cup walls are usually thicker and have been strain hardened more than the bottom, so the punch force actually increases during the stroke; late failures are likely. In contrast to cupping, the limiting reduction in redrawing should decrease with increasing n-values because strain hardening increases the drawing load near the end of

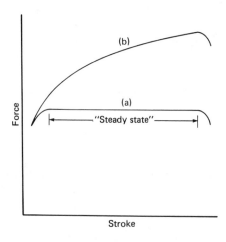

(b)

(a)

"Steady state"

Force

Stroke

Figure 14-14   Variation of punch force during redrawing. For a non-work hardening metal of constant wall thickness (a), the force is constant until the amount of material in the flange decreases. For a work-hardening metal (b), initial cupping causes more work hardening high in the walls, so the redraw force rises with stroke. Increased wall thickness has a similar effect.

the stroke without appreciably strengthening the bottom of the walls. Thus, the limiting redraw ratio is greater with materials of low n-value (e.g., cold-rolled sheets) than with high n-value (e.g., annealed sheets). High $\bar{R}$-values are beneficial, since reduction of the wall diameter corresponds to flange deformation in cupping (i.e., plane strain with $\epsilon_z = 0$) and failure by wall necking involves $\epsilon_y = 0$, plane strain. Therefore, large values of $\bar{R}$ or β permit greater reductions.

14-8 IRONING

If ironing (wall thinning) is employed during cupping or redrawing, a more uniform wall thickness and increased cup height result. Often, only the top of the wall is ironed, this being achieved by controlling the clearance between the punch and the die lip, so no ironing occurs until the thickened material on the outer part of the flange flows over the die lip. Ironing adds to the total punch force, but if it occurs only late in the draw after the maximum punch force is reached, it does not affect drawability.† This is shown schematically in Fig. 14-15.

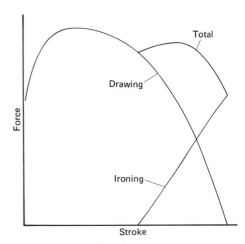

Figure 14-15 Variation of drawing force with stroke when ironing occurs late in the draw. Drawability will be decreased if so much ironing occurs that the second maximum is greater than the first.

Ironing decreases the degree of earing, besides producing a thinner and more uniform wall thickness. Since the wall thickness in the valleys between ears is greater than at the eared portions, the increase in height due to ironing is greater at the valleys. A more uniform height (i.e., less pronounced earing) results.

Sometimes severe ironing is accomplished in separate stages on drawn or redrawn cups to effect appreciable increases in wall height, as in the manufacture of two-piece beer cans where the walls are about one third the thickness of the original sheet. The flow in such ironing can be viewed as plane-strain drawing and treated by slip-line field analysis. There is, of course, a limiting reduction for a single ironing stage. With higher reductions, the ironing forces exceed the wall strength. As with redrawing, the maximum ironing reductions decrease with work-hardening capacity (higher n-values) because the bottom of the wall is not strengthened as much in the initial cupping (and redrawing) as positions higher on the wall; see Fig. 14-16. Again, heavily cold-worked sheets will outperform annealed ones.

Unlike cupping and redrawing, ironing does not benefit from high R

†Excessive ironing may cause a second maximum in the drawing force, thereby reducing the LDR. This is not the usual case.

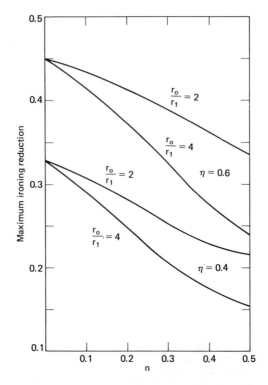

Figure 14-16 Calculated dependence of maximum ironing reduction per pass upon strain-hardening exponent, n, efficiency, η, and prior drawing and redrawing reduction, r_0/r_1, where r_0 is the initial blank radius and r_1 is the radius of the cup being ironed. From W. F. Hosford, *ibid.*

values. This is because ironing is characterized by the same plane-strain flow, $d\epsilon_y = 0$, as the potential necking failure of the wall. The only influence of the R-value is its effect on wall thickness changes before ironing. With high R-values, there is less wall thickening in the initial drawing and redrawing steps, so smaller ironing reductions are required to achieve the same wall thickness.

A drawing, redrawing, and ironing sequence can be achieved with a single stroke of concentric punches as shown schematically in Fig. 14-17.

14-9 RESIDUAL STRESSES

There are large residual stresses left in cups after drawing. In the axial direction there is residual tension on the outside and compression on the inside, resulting from bending and unbending over the die lip. These stresses are largest near the top of the wall because of the bending that occurred near the end of the draw, when there was little net tension. These residual stresses induce a bending moment in the wall which is balanced by hoop tension near the top of the cup. In some

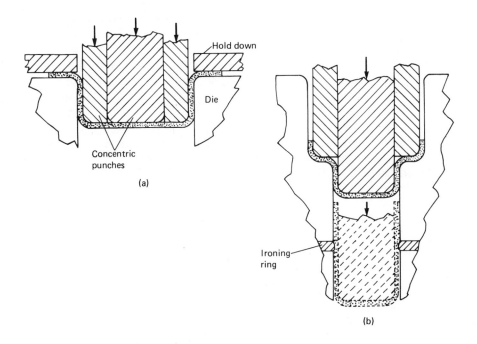

Figure 14-17 Schematic illustration of tooling for making deep cups using a continuous stroke with concentric punches; (a) the cupping stage and (b) redrawing and ironing stages.

materials this hoop tension can lead to splitting of the walls by stress-corrosion cracking (see Fig. 14-18) unless the cups are stress relieved. Other metals, especially ones with aligned inclusions, crack immediately as the cup leaves the die. Ironing, which causes the entire section to yield, greatly diminishes the magnitude of the residual stresses.

Figure 14-18 Stainless steel cups which failed by stress corrosion cracking in a laboratory atmosphere within 24 hours after drawing. The cracks are perpendicular to the residual hoop tension in the walls. From D. J. Meuleman, *ibid.*

PROBLEMS

14-1. Calculate the height-to-diameter ratios for cups drawn with drawing ratios of 1.8, 2, 2.25, and 2.5. Assume constant thickness.

14-2. Calculate the slope of the LDR versus $\bar{R}$ curve at $\bar{R} = 1$ and $\eta = 0.75$ according to:

(a) Whiteley's equation, Eq. (14-14).

(b) Whiteley's analysis using Eq. (14-15).

(c) Graphically from Fig. 14-5 by drawing the best curves through the data.

14-3. **(a)** Derive an expression relating R to β from Eqs. (13-18) and (13-19).

    (b) What value of the exponent, a, in Eqs. (13-18) and (13-19) gives the same results for the slopes of the LDR versus R curves as found in Prob. 14-2c?

14-4. Show that the frictional force acting on the die lip is approximately equal to $\pi\mu F_d/2$. (*Hint:* Make a force balance on a differential segment on the die lip and integrate between the limits $F = F_f$ at $\theta = 0$ and $F = F_d$ at $\theta = \pi/2$; the "frictional force" is $F_d - F_f$.)

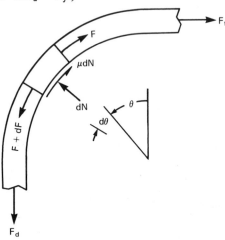

14-5. Prove Eq. (14-11).

14-6. A typical aluminum beer can is $5\frac{1}{4}$ in. high and $2\frac{7}{16}$ in. in diameter, with a wall thickness of 0.005 in. and thickness on the bottom of 0.016 in. It is made from sheet by drawing, redrawing, and ironing. The starting material is 3004-H19 (already cold rolled in excess of 80%).

    (a) What diameter blank is required?

    (b) Is a redrawing step required? (Assume that a safe drawing ratio is 1.8.) Are multiple redraws required? If so, how many?

    (c) How many ironing stages would be required? (Assume a deformation efficiency of about 50%.)

    (d) If this operation were carried out with one continuous stroke of telescoping punches, and if each operation must be completed before the next one starts, how long must the punch stroke be?

14-7. In the analysis of deep drawing, the work in bending and unbending as the sheet flows over the die lip was neglected.

    (a) Derive an expression for the contribution of the energy expended in bending and unbending to the drawing force in terms of the average strain ratio, $\bar{R}$, the x-direction yield stress, X, the sheet thickness, t, and the radius of curvature of the die lip, r. Assume radial isotropy.

    (b) What fraction of the total drawing load might come from this source for a sheet with $R = 1$, $t = 0.032$ in., $r = 0.250$ in., $d_o = 2.2$ in., $d_1 = 1.0$ in.?

14-8. It has been claimed that significant increases in limiting drawing ratio can be achieved by using a pressurized fluid (e.g., water) on the punch side of the blank as shown in the sketch. Limiting drawing ratios up to 3.0 are claimed for materials like low-carbon steel that would, with conventional tooling, have an LDR of about 2.1.

(a) Explain the basis for these claims. That is, describe how the pressurized fluid acts to increase drawability.

(b) Estimate the level of pressure required for LDR = 3.0. Assume isotropy, no work hardening, $Y = 40$ ksi, and $\eta = 0.75$.

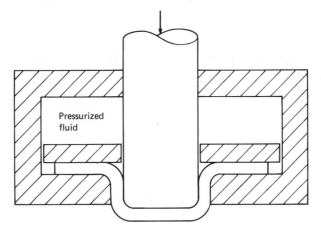

Complex stampings

15-1 LOCALIZED NECKING IN BIAXIAL STRETCHING

In many sheet-forming operations, the deformation is characterized by biaxial stretching instead of drawing, as discussed in Chap. 14. Failure in stretching operations normally occurs by the development of a sharp localized neck on the surface. Localized necking should not be confused with diffuse necking, which precedes it and normally leads to failure in tension tests of round bars. To understand the distinction between a localized neck and a diffuse neck, first consider the deformation of a wide, thin sheet specimen loaded in uniaxial tension in the 1-direction.

Initially the deformation is uniform. When the load reaches a maximum, a diffuse neck starts to form (Fig. 15-1a). For a material obeying power-law hardening, $\bar{\sigma} = K\bar{\epsilon}^n$, the strain at the onset of diffuse necking will be $\epsilon_1^* = n$, as derived earlier. (The * indicates the critical strain at instability). The diffuse neck is accompanied by contraction strains in both the width and thickness directions, and for an isotropic material, $d\epsilon_2 = d\epsilon_3 = -\frac{1}{2}d\epsilon_1$. With a wide specimen, the width strain, $d\epsilon_2$, cannot localize rapidly, so the whole neck develops gradually and considerable extension is still possible after the onset of diffuse necking. A condition will finally be reached where a sharp localized neck can form at an angle, θ, to the loading axis (Fig. 15-1). Typically, the width, b, of the neck is of the order of the sheet thickness, t, so very little additional elongation is possible before failure.

The characteristic angle, θ, of the neck and the strain at which it forms depend upon the geometry. Since the neck is very narrow, the strain parallel

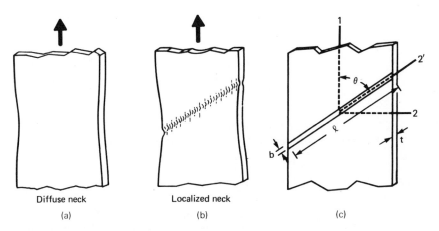

Figure 15-1 Development of a diffuse neck (a) and a localized neck (b). Coordinate axes used in analysis are shown in (c).

to the neck, $d\epsilon_{2'}$ must be zero. However, $d\epsilon_{2'}$ is directly related to $d\epsilon_1$ and $d\epsilon_2$ by the strain transformation, Eq. (1-34), so

$$d\epsilon_{2'} = d\epsilon_1 \cos^2\theta + d\epsilon_2 \sin^2\theta = 0 \qquad (15\text{-}1)$$

For uniaxial tension in the 1-direction, the flow rules for an isotropic material predict that

$$d\epsilon_2 = -\frac{d\epsilon_1}{2} = d\epsilon_3 \qquad (15\text{-}2)$$

Substituting Eq. (15-2) into Eq. (15-1),

$$d\epsilon_1 \cos^2\theta + \left(-\frac{d\epsilon_1}{2}\right)\sin^2\theta = 0$$

so $\qquad \dfrac{\sin^2\theta}{\cos^2\theta} = 2 \quad$ or $\quad \tan\theta = \sqrt{2} \quad$ and $\quad \theta = 54°44'$ $\qquad (15\text{-}3)†$

If the metal is anisotropic, $d\epsilon_2/d\epsilon_3 = R$, or $d\epsilon_2/d\epsilon_1 = -R/(R+1)$. In this case,

$$\theta = \arctan\left(\sqrt{(R+1)/R}\right) \qquad (15\text{-}4)$$

The cross-sectional area of the neck, A', equals ℓt. Since $A' = \ell t$ and ℓ is constant, $dA'/A' = dt/t = d\epsilon_3$. The area perpendicular to the 1-axis is $A = A' \sin\theta$, but θ is also constant so

$$\frac{dA}{A} = \frac{dA'}{A'} = d\epsilon_3 \qquad (15\text{-}5)$$

The local neck can form only if the load, F, can fall under the constraint, $d\epsilon_{2'} = 0$. Since $F = \sigma_1 A$,

$$dF = 0 = \sigma_1 dA + A d\sigma_1 \qquad (15\text{-}6)$$

†This can also be shown by plotting the Mohr's circle of stress and corresponding plastic strain.

or
$$\frac{d\sigma_1}{\sigma_1} = \frac{-dA}{A} = -d\epsilon_3 \tag{15-7}$$

but
$$d\epsilon_3 = \frac{-d\epsilon_1}{2}$$

so
$$\frac{d\sigma_1}{\sigma_1} = \frac{d\epsilon_1}{2} \tag{15-8}$$

If $\bar{\sigma} = K\bar{\epsilon}^n$, then $\sigma_1 = K\epsilon_1^n$ for uniaxial tension, and $d\sigma_1 = nK\epsilon_1^{n-1} d\epsilon_1$. Therefore, the critical strain for localized necking in uniaxial tension becomes

$$\epsilon_1^* = 2n \tag{15-9}$$

while, as indicated earlier for diffuse necking, $\epsilon_1^* = n$.

In sheet-metal forming, the stress state is rarely one of uniaxial tension, and the state of strain is rarely that $d\epsilon_2 = d\epsilon_3 = -d\epsilon_1/2$. However, the same principles can be used to develop the conditions for localized necking under a general state of biaxial tension. Assume for now that the loading during stretching maintains a constant strain ratio,† $\rho = \epsilon_2/\epsilon_1$. Substitution of $\epsilon_2 = \rho\epsilon_1$ into Eq. (15-1) gives

$$\epsilon_1 \cos^2\theta + \rho\epsilon_1 \sin^2\theta = 0 \tag{15-10}$$

or
$$\tan\theta = \frac{1}{\sqrt{-\rho}} \tag{15-11}$$

The angle θ will have a real value only if ρ is negative (i.e., ϵ_2 is negative). Physically this means that if ϵ_2 is positive, there is no angle at which a local neck can form and still fulfill the requirement that $\epsilon_{2'} = 0$ (i.e., if ρ is constant and > 0, local necks cannot form). As the stress becomes more biaxial, α increases and ρ becomes less negative, so θ must increase. For plane-strain conditions ($\epsilon_2 = 0$), $\rho = 0$, so $\tan\theta = \infty$ and $\theta = 90°$.

The critical strain for necking is also influenced by ρ. From constancy of volume,

$$d\epsilon_3 = -(1 + \rho)\, d\epsilon_1 \tag{15-12}$$

so Eq. (15-7) becomes

$$\frac{d\sigma_1}{\sigma_1} = (1 + \rho)\, d\epsilon_1 \tag{15-13}$$

Using $\sigma_1 = K\epsilon_1^n$, the condition for necking becomes‡

$$\epsilon_1^* = \frac{n}{1 + \rho} \tag{15-14}$$

Equation (15-14) implies that the critical strain, ϵ_1^*, for local necking decreases from $2n$ for $\rho = -\frac{1}{2}$ to n for plane strain (i.e., $\rho = 0$).

---

†Note that this is equivalent to assuming a constant stress ratio $\alpha = \sigma_2/\sigma_1$. For isotropy, the flow rules with $\sigma_3 = 0$ give $\rho = \epsilon_2/\epsilon_1 = (\sigma_2 - \sigma_1/2)/(\sigma_1 - \sigma_2/2)$. With substitution of $\sigma_2 = \alpha\sigma_1$, the relationship between α and ρ can be expressed as $\rho = (2\alpha - 1)/(2 - \alpha)$, or $\alpha = (2\rho + 1)/(2 + \rho)$.

‡Strictly the power law should be written as $\bar{\sigma} = K\bar{\epsilon}^n$, where $\bar{\sigma}$ is the effective stress and $\bar{\epsilon}$ is the effective strain, but a more rigorous derivation gives the same result.

Swift† has shown that diffuse necking can be expected when

$$\epsilon_1^* = \frac{2n(1 + \rho + \rho^2)}{(\rho + 1)(2\rho^2 - \rho + 2)} \tag{15-15}$$

The criteria for local necking and diffuse necking given by Eqs. (15-14) and (15-15) are plotted in Fig. 15-2.

The previous analysis seems to imply that localized necks cannot form under stretching conditions where ϵ_2 is positive. Indeed, if loading could be maintained with constant α (a constant ρ), no localized necking can occur and

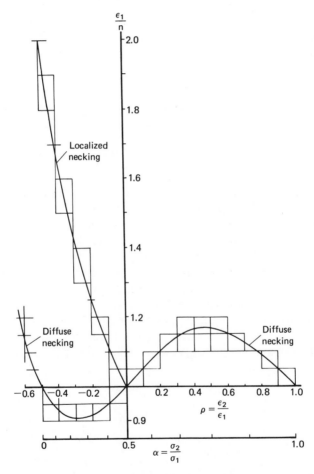

Figure 15-2 Critical strains for diffuse and localized necking according to Eqs. (15-14) and (15-15). Loading with constant stress and strain ratios is assumed. Note that under these conditions localized necking cannot occur if ϵ_2 is positive.

†H. W. Swift, "Plastic Instability Under Plane-Stress," *J. Mech. and Phys. of Sol.*, 1 (1952), pp. 1–18.

stretching would continue until fracture occurred. However, α and p do change during stretching. Local necks can be explained by changing strain paths even though the total strain $\epsilon_2 > 0$. What is critical for the formation of a local neck, is that the incremental ratio, $p' = d\epsilon_2/d\epsilon_1$ becomes zero rather than the total strain ratio $p = \epsilon_2/\epsilon_1$. It has been argued[†] that because of local inhomogeneities in the material (variations of grain size, texture, alloying elements or sheet thickness), there can exist a zone of weakness, or trough, lying parallel to the 2-direction (Fig. 15-3). Although such a trough is not a true neck, it may develop into one. The strain ϵ_1 will grow faster in the weak trough than outside of it. However, the strain ϵ_2 must be the same in the trough as on either side of it because of constraint by the surrounding material. Therefore, within the trough ϵ_1 increases faster than ϵ_2, so the local value of $p' = d\epsilon_2/d\epsilon_1$, decreases. Once p' reaches zero, a true local neck can form.

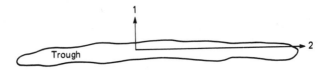

Figure 15-3   Sketch of a trough parallel to the 2-axis. The material in the trough is either thinner or weaker than material outside it.

Often the tool geometry itself will induce changes in the loading path so that postulated material inhomogeneities are not necessary to explain why $d\epsilon_2/d\epsilon_1$ decreases during a forming operation. Consider, for example, a sheet being stretched over a rough hemispherical dome (see Fig. 15-4). The flange is locked to prevent any drawing. If friction between the punch and the sheet is sufficient to prevent sliding, deformation will cease in an element once it contacts the punch. An element in the free section between the die and punch is stretched biaxially; the strains in both the radial direction, ϵ_1, and in the hoop (circumferential) direction, ϵ_2, being positive. As an element approaches contact with the punch, the rate of increase of ϵ_2 will lessen because of circumferential constraint by neighboring elements on the punch; the radial strain, ϵ_1, will not

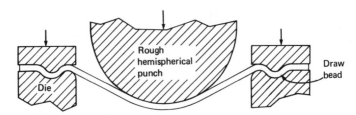

Figure 15-4   Sketch of a rough hemispherical punch stretching a sheet; sticking friction is assumed.

†Z. Marciniak, *Archiwum Mechanikj Stosowanej*, 4 (1965), p. 579; and also, Z. Marciniak and K. Kuczynski, *Int. J. Mech. Sci.*, 9 (1967), p. 609.

be so constrained. As an element makes contact with the punch, $d\epsilon_2/d\epsilon_1 \rightarrow 0$ so that necking can occur. In actual tests of this sort, failure does occur by plane-strain necking along the ring of punch contact, and with lower friction, the positions of no sliding and failure are moved toward the center of the dome.

15-2 FORMING LIMIT DIAGRAMS

The above discussion explains qualitatively how local necks can occur under biaxial stretching ($\epsilon_2 > 0$). Quantitative analyses are complex and of questionable accuracy. However, the strains ϵ_1^* at which local necks are first observed have been experimentally measured for various materials and loading paths. The most widely used technique involves printing or etching a grid of small circles of diameter d_o on the metal sheet before forming; during forming these circles are distorted into ellipses. The principal strains can be determined by measuring the major diameter, d_1, and the minor diameter, d_2. By convention, *engineering strains* $e_1 = (d_1 - d_o)/d_o$ and $e_2 = (d_2 - d_o)/d_o$ have been reported. These values at a neck or fracture give the "failure" condition, while strains in circles one or more diameters from a failure are considered "safe" (see Fig. 15-5). By plotting these measured strains it is possible to construct a forming limit diagram (FLD), or Keeler-Goodwin diagram. A typical FLD for a low-carbon steel is shown in Fig. 15-6, and Fig. 15-7 is a comparison of this FLD with the theoretical conditions for diffuse and local necking as expressed by Eqs. (15-14) and (15-15). For negative values of e_2, the experimental curve parallels the theoretical curve for localized necking. The fact that it is higher probably indicates that some of the "safe" experimental values were affected by nearby necks. The lowest value of e_1 occurs at plane-strain conditions ($e_2 = 0$), e_1 rising again for biaxial straining.

The whole level of the forming limit diagram rises with both the strain-hardening exponent, n, and the sheet thickness, t. This is illustrated for low-carbon steels in Fig. 15-8, where FLD_o is the level of e_1^* at plane strain, $e_2 = 0$. To a first approximation, increasing t displaces the entire forming limit curve upward. Increasing n has the same effect, so the FLD for low-carbon steel can be estimated from n and t.

Forming limit diagrams for several other materials are shown in Fig. 15-9; the different levels of the curves are not completely explained by differing values of n and t. The relatively high level of the FLD for low-carbon steel is due in part to its higher strain-rate sensitivity.

Forming limit diagrams have proved to be very useful for diagnosing actual and potential problems in sheet-forming operations.[†] Sheets premarked with circles are formed in either prototype or production tools. Local strains near failures or suspected trouble points are measured and compared with the

†See S. P. Keeler, "Understanding Sheet Metal Formability," Parts IV and VI, *Machinery Magazine* (May and July 1968).

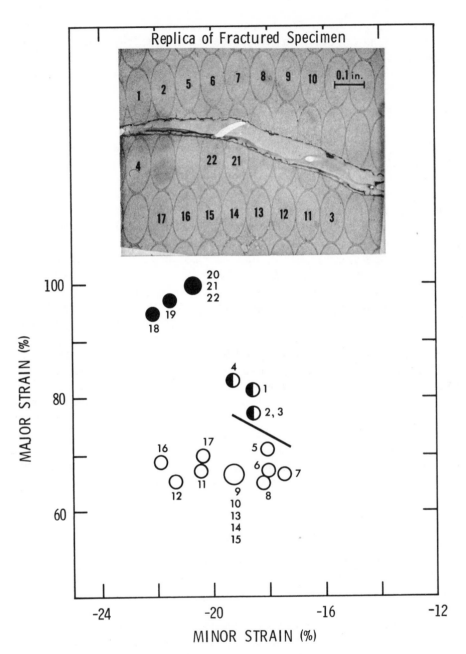

Figure 15-5 Distortion of printed circles near a localized neck and a plot of the strains in the circles. Solid points are for grid circles through which failure occurred, open points are for grid circles removed from failure, and partially filled points are for grid circles very near failure. From S. S. Hecker, *Sheet Metal Ind.*, 52 (1975), pp. 671–75.

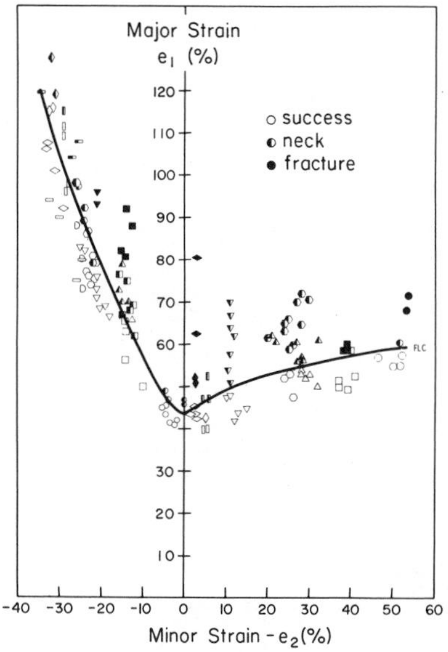

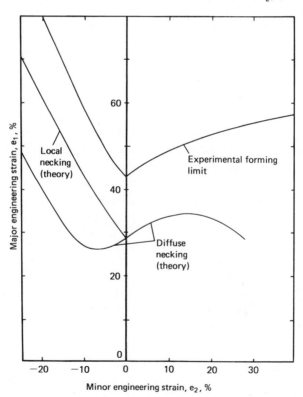

Figure 15-6 Forming limit diagram for low-carbon steel determined from data like that in Fig. 15-5. The strains below the curve are acceptable while those above the curve correspond to regions affected by local necking. From S. S. Hecker, *ibid.*

Figure 15-7 Comparison of the experimental FLD in Fig. 15-6 with the theoretical curves of Fig. 15-2, replotted as engineering strain, using $n = 0.25$.

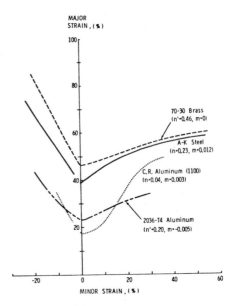

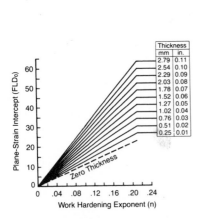

MAJOR
STRAIN, (%)

70-30 Brass
(n'=0.46, m=0)

A-K Steel
(n=0.23, m=0.012)

C.R. Aluminum (1100)
(n=0.04, m=0.003)

2036-T4 Aluminum
(n'=0.20, m=-0.005)

MINOR STRAIN, (%)

Thickness	
mm	in.
2.79	0.11
2.54	0.10
2.29	0.09
2.03	0.08
1.78	0.07
1.52	0.06
1.27	0.05
1.02	0.04
0.76	0.03
0.51	0.02
0.25	0.01

Zero Thickness

Work Hardening Exponent (n)

Figure 15-8 Effect of the strain-hardening exponent, n, and sheet thickness on the forming limit diagram. FLD_0 is the level of e_1 for plane strain, $\epsilon_2 = 0$. From S. P. Keeler and W. G. Brazier, in *Micro Alloying 75*, Union Carbide, N.Y. (1977), pp. 517-30.

Figure 15-9 Forming limit diagrams for several materials. From A. K. Ghosh, *J. of Eng. Matls. and Tech., Trans. ASME*, Series H, 99 (1977) pp. 264-74.

FLD; this serves two useful purposes. First, potential trouble spots can be identified and the severity of strain assessed even though failure did not occur. If the strains are near the failure curve, problems are likely to occur in production because of tool wear, variations in lubrication, tool alignment, and material thickness and properties. (In production one failure in 20 stampings causes severe problems.) The second advantage in comparing strains with the FLD is that the type of problem can be diagnosed. The lowest value of ϵ_1^* occurs at plane strain ($\epsilon_2 = 0$), so if the deformation in a critical region is nearly plane strain, tooling or lubrication changes which induce either more drawing or more biaxial stretching will be beneficial. Better lubrication and less flange locking often can be used to promote drawing. Steel vendors often test gridded sheets in dies to determine the quality of steel required to assure a low failure rate.

15-3 STRAIN DISTRIBUTION

Forming limit diagrams alone do not completely describe the forming behavior of different materials. Two materials may have nearly the same forming limits (and hence failure strains e_1^*), but may differ substantially in forming behavior.

For example, consider a symmetric part with a line of length L_o scribed on the blank parallel to the 1-axis. As the punch descends to a depth, h, the length of the line will be increased to L and the part depth, h, will depend upon L [i.e., $h = f(L/L_o)$]. The stress in the sheet is not uniform and therefore the values of e_1 vary with position, x, along the line. However, the total length change can be expressed as

$$\frac{L}{L_o} = \frac{1}{L_o} \int_O^{L_o} (1 + e_1)\, dx \tag{15-16}$$

The height at failure, h_{max}, depends upon the area under a curve of e_1 versus x at the time the forming limit is reached ($e_1 = e_1^*$). That is, h_{max} is strongly influenced by the distribution of e_1 versus x. The ratio of e_1 in a lightly loaded region to that in a heavily loaded region depends primarily on n, but also to some extent on m. This is analogous to the derivation in Chap. 4 which treated the relative strains in two regions of a stepped tensile specimen. Increasing values of n and m tend to distribute the strain more widely and permit deeper parts to be formed. This is illustrated in Fig. 15-10.

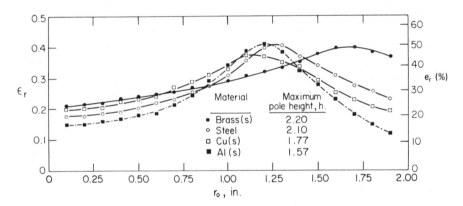

Figure 15-10   Distribution of strains in sheet stretching for several materials. The punch was stopped at the same peak strain, but the cup depth increases as the strain is more widely distributed. Brass has the highest n and aluminum the lowest. From S. P. Keeler and W. A. Backofen, *Trans. Q. ASM*, 56 (1963), pp. 25–48.

15-4 CUPPING TESTS

The Erichsen and Olsen cupping tests have been widely used to assess sheet formability. In both tests the sheet is clamped between two polished flat plates with a hole of diameter, D, and a ball of diameter, d, is pressed into the sheet until failure occurs (see Fig. 15-11). The height of the cup, h, is used as the formability index. These tests are now losing favor because of irreproducibility of data and lack of correlation with either other properties or service experience.

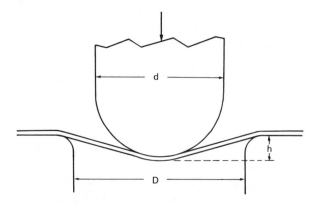

Figure 15-11. Erichsen and Olsen cup tests.

Hecker† attributes this to "insufficient size of the penetrator, inability to prevent inadvertent drawing-in of the flange, and inconsistent lubrication." Instead he has proposed a test in which drawbeads prevent drawing-in. The penetrator is a hemispherical punch of diameter 101.60 mm and the hole diameter is 105.66 mm. The test is run dry (without lubrication) to simulate better most press operations. For this test the limiting cup height depends upon both the n- and m-values of the sheet, as shown in Fig. 15-12. The correlation with total elongation in tension is shown in Fig. 15-13. This correlation may at first seem surprising because total elongation is not a fundamental quantity. Total elonga-

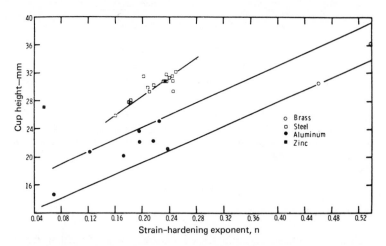

Figure 15-12 Maximum cup height for stretching over a dome as a function of n for various materials. The corresponding m values are: steel 0.01, brass 0.0, zinc 0.052, and aluminum 0.0 to -0.006. From S. S. Hecker, *Met. Eng. Q.*, 14 (1974), pp. 30–36.

†S. Hecker, *Met. Eng. Q.*, 14, no. 4 (1974), p. 30.

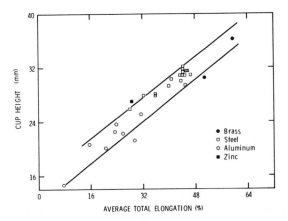

Figure 15-13 The same data on maximum cup height for biaxial stretching as shown in Fig. 15-12, plotted as a function of average total elongation in tension, $\bar{e} = (e_o + 2e_{45} + e_{90})/4$. The improved correlation occurs because both total elongation and cup stretchability depend on m as well as n. From S. S. Hecker, *ibid.*

tion includes both the uniform elongation and the post-necking elongation. (Therefore it depends upon the length, width, and thickness of the gauge section, and these must be standardized for meaningful comparisons of different materials.)

The reason that *total* elongation is important is that in many stamping operations, useful stretching occurs after the onset of diffuse necking, and the amount of such post-necking strain depends upon the same properties as does the post-necking strain in a tension test. This can be seen by comparing the stress-strain curves in Fig. 15-14 with the post-necking elongation (Fig. 15-15). It is apparent that the strain-rate sensitivity, m, plays an important role in governing both total elongation and formability.

15-5 EDGE CRACKING

Formability is not always limited by localized necks. Sometimes cracking occurs at sheared edges where the edges are elongated in tension or subjected to sharp bends. The tendency to edge cracking is greatly aggravated if shearing has left a burred edge.† Therefore, sharp and aligned tools decrease the problem. Edge-cracking problems vary greatly with material; they are particularly severe in high-strength steels with elongated inclusions. The sensitivity to edge cracking is often assessed by the hole-expansion test in which a punched hole is enlarged with a conical or hemispherical punch until cracking is observed; the circumferential strain at cracking, $e_1 = (D - D_o)/D_o$, is measured.

†S. P. Keeler, *Machinery Magazine* (June 1968).

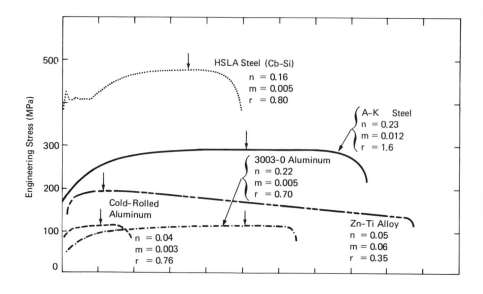

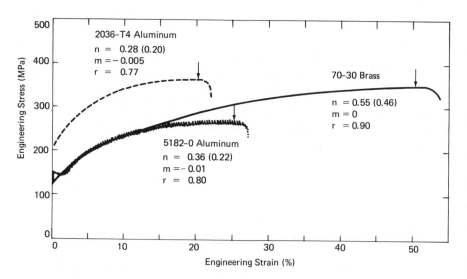

Figure 15-14 Engineering stress strain curves showing that the post-necking elongation depends on *m*. Vertical arrows indicate maximum load. From A. K. Ghosh, *ibid.*

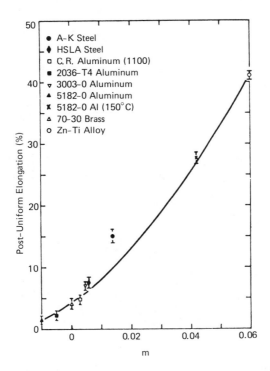

Legend (in figure):
- A-K Steel
- HSLA Steel
- C.R. Aluminum (1100)
- 2036-T4 Aluminum
- 3003-0 Aluminum
- 5182-0 Aluminum
- 5182-0 Al (150°C)
- 70-30 Brass
- Zn-Ti Alloy

Y-axis: Post-Uniform Elongation (%)

X-axis: m

Figure 15-15 Post-necking elongation in this figure correlates well with m. From A. K. Ghosh, *ibid.*

15-6 SPRINGBACK

Control of springback is often of great importance, particularly in lightly curved sections like the bottom of the stamping shown in Fig. 15-16. As indicated in Chap. 12, springback will be minimized if the tensile force, F_1, in the bottom is sufficient to cause yielding throughout the cross section. Yet the wall force, F_2, must not cause a stress that exceeds the tensile strength. The relative magnitudes of F_1 and F_2 can be calculated approximately as follows: a radial force balance on an element in the bend gives the normal force $dN = Fd\theta$, so the frictional force is $\mu dN = \mu F d\theta$, where μ is the coefficient of friction. From a force balance in the circumferential direction,

$$F + dF = F + \mu dN = F + F\mu d\theta$$

or

$$\int_{F_1}^{F_2} \frac{dF}{F} = \mu \int_0^\theta d\theta \quad \text{so} \quad F_2 = F_1 e^{\mu\theta} \tag{15-17}$$

Neglecting the difference between the plane strain and uniaxial values of tensile and yield stress, $F_2 < (S_u)wt$ and $F_1 > (Y)wt$, where w is the dimension parallel

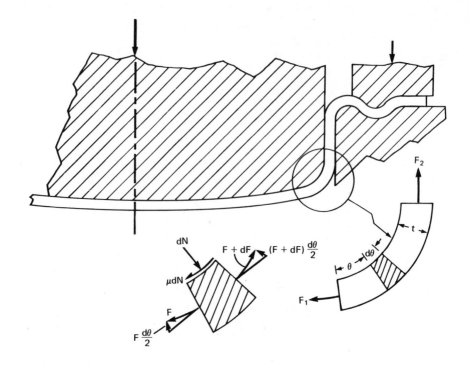

Figure 15-16 Schematic drawing for analysis of friction on tensile stresses during stamping of a lightly curved panel.

to the bend, t is the sheet thickness, S_u is the tensile strength, and Y is the yield strength. From Eq. (15-17)

$$\frac{S_u}{Y} > e^{\mu\theta} \qquad (15\text{-}18)†$$

Thus, the ratio of tensile-to-yield strength must exceed a value which depends upon the friction coefficient and bend angle; otherwise, the bottom cannot yield without the walls failing. This can cause problems in the forming of high-strength steels and aluminum alloys which have low ratios of tensile to yield strengths. For example, if $\mu = 0.2$ and $\theta = \pi/2$, $S_u/Y > 1.37$.

15-7 GENERAL OBSERVATIONS

The relative amounts of drawing and stretching vary in stampings. Where stretching predominates, formability depends mainly on n and m, whereas if drawing predominates, $\bar{R}$ is most important. Figure 15-17 relates the press performance of low-carbon steels in three different automobile components

†This analysis was pointed out to the authors by J.L. Duncan.

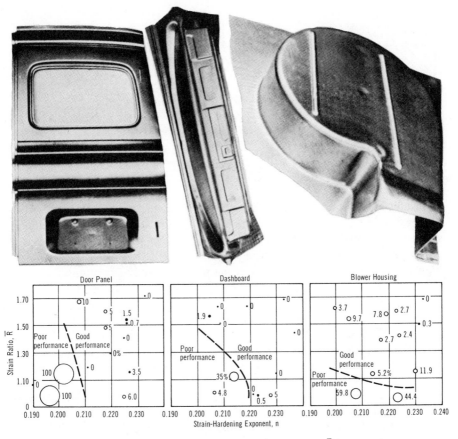

Figure 15-17 Dependence of forming behavior upon n and $\bar{R}$ for several automobile parts. Circles and numbers indicate percentage failures. When stretching predominates (left), n is most important and where drawing predominates (right), $\bar{R}$ is most important. From D. J. Blickwede, *Metals Prog.* (December 1968), pp. 64–70.

to n and $\bar{R}$. Here the role of m is not shown, but m would not vary significantly in these steels. However, in comparisons between steel and aluminum alloys, the higher values of m for steel often result in much better press performance.

In the stamping of a given part, the relative amount of drawing and stretching is affected significantly by die design. Draw beads are used to decrease the amount of drawing-in of the flange because excessive drawing causes wrinkling of unsupported regions.

The tendency to wrinkle depends on the R-value in some cases. During drawing of a conical cup, Fig. 15-18, wrinkling may occur in the unsupported walls. Consider an annular element at A which was initially at A'. Its circumference must decrease, as it is elongated in the radial direction. For a material with a low R-value, the contractile strain in the circumferential direction may

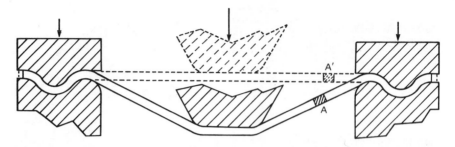

Figure 15-18 Schematic illustrating the possibility of wrinkling in drawing a conical cup. Note that as an element at A' moves to A, its circumference decreases.

be insufficient to accommodate the decrease in circumference; in this case, wrinkling must occur. For a material with a high R-value, the larger circumferential strain would prevent wrinkling. Although the circumferential shrinkage at A could be minimized by using stronger draw beads to prevent drawing-in, the higher radial strain necessary to form the part may lead to failure.

PROBLEMS

15-1. Derive an expression for the critical strain to produce diffuse necking, ϵ_1^*, as a function of n and the stress ratio, α. Assume the von Mises criterion and loading under constant α. [*Hint:* start with Eq. (15-15)].

15-2. In principle, one can determine the R-value by measuring the characteristic angle, θ, of a neck in a strip tension test. How accurately would one have to measure θ to distinguish between two materials having R-values of 1.6 and 1.8?

15-3. It has been suggested that failure in sheet metals occurs when the thickness strain reaches a critical value. Superimpose a plot of e_1 vs. e_2 for this criterion on Fig. 15-7. Adjust the constant so that the curves match at plane-strain conditions ($e_2 = 0$). How good is this suggestion?

15-4. Repeat 15-3 for a failure criterion that predicts failure when the absolutely largest strain, $|\epsilon_i|_{max}$, reaches a critical value.

15-5. Ironing of the walls of a two-piece aluminum beverage can reduces the wall thickness from 0.015 in. to 0.005 in. Compare this deformation with the forming limit curves for aluminum alloys (Fig. 15-9) and explain why failure is not encountered in the ironing.

15-6. Coefficients of friction in metal-forming operations are significantly higher than those measured in sliding experiments because new surface is being formed. Figure problem 15-6 illustrates a method proposed to measure the coefficient of friction under conditions similar to the sliding of metal over a die punch lip. A strip of metal is bent around a fixture consisting of two non-rotating cylinders (which simulate tools) and is stretched in tension. Tensile strains are measured in regions A and B. For a metal with a strain-hardening exponent of $n = 0.22$, $\epsilon_B = 0.040$ when $\epsilon_A = 0.180$. Calculate the coefficient of friction on the cylinders.

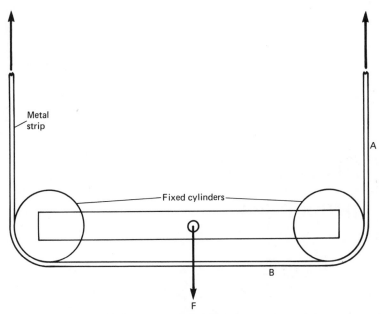

15-7. Consider a bulge test on a sheet which is clamped at the periphery over a circular orifice of radius, R. The sheet is bulged by hydraulic pressure until the center has a height, h, such that $h/R = 0.5$. See the figure.

For the sake of analysis, assume the bulged shape has a constant radius of curvature so that it is a segment of a sphere, and also assume that the radial strain is constant everywhere. (Neither of these assumptions is really correct.)

(a) Calculate the radial strain, ϵ_r.

(b) Calculate the circumferential strain, ϵ_c, at points which were initially at various radii, r_o, from the center.

(c) Comparing ϵ_c and ϵ_r, deduce how the stress ratio, α, varies with position from the center.

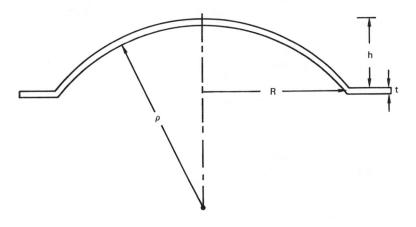

16

Sheet metal
properties

16-1 INTRODUCTION

The properties important in sheet forming tend to vary from one class of materials to another, but some generalization can be made. Table 16-1 gives typical ranges of values of strain-hardening exponent, n, average strain ratio, $\bar{R}$, and strain-rate senstitivity, m, for several classes of annealed alloys. It should be

TABLE 16-1   Typical Sheet Metal Properties†

	n	$\bar{R}$	m
L.C. steel (rimmed)	0.20–0.25	1.0–1.4	0.01–0.015
L.C. steel (killed)	0.22–0.26	1.4–1.8	0.015
Interstitial-free steel	0.30	1.8–2.5	0.015
HSLA steels	0.10–0.18	0.9–1.2	0.005–0.01
Stainless steel (ferritic)	0.16–0.23	1.0–1.2	0.01–0.015
Stainless steel (austenitic)	0.40–0.55	0.9–1.0	0.01–0.015
Copper	0.35–0.50	0.6–0.9	0.005
Brass (70-30)	0.45–0.60	0.8–0.9	0.0–0.005
Aluminum alloys	0.20–0.30	0.6–0.8	−0.005–+0.005
Zinc alloys	0.05–0.15	0.4–0.6	0.05–0.08
α-titanium alloys	0.05	3.0–5.0	0.01–0.02

†Although these values are typical, there is variation from supplier to supplier and from lot to lot depending upon composition, rolling, and annealing practice. Values higher or lower than those quoted here may be obtained. In general, as strength levels are increased by precipitation or grain-size refinement, the levels of n and m tend to fall.

noted, however, that while these values are typical, higher or lower values can arise from variations in composition or rolling and heat-treating practices.

16-2 SURFACE APPEARANCE

Surface appearance is often of great concern when selecting a material for a drawing application. Several types of surface defects may be produced during sheet forming. One is an *orange peel* effect, which is a surface roughness of about the scale of the grain size (see Fig. 16-1). Because neighboring grains on the surface have different orientations, the different tendencies to thicken or thin during deformation result in a roughened surface. Orange peel is observed only if there is a free surface not in intimate contact with tools. The most obvious way of diminishing the extent of orange peel is to use a material of fine enough grain size so that any resulting roughness is either not apparent or on so fine a scale that it is not seen after painting or plating. A related surface phenomenon is the *ridging* or *roping* found in ferritic stainless steels. These steels are characterized by duplex textures; that is, most of the grains have orientations close to either of two ideal orientations. Similarly oriented grains tend to be clustered together and the clusters are elongated in the rolling direction. As the sheet is stretched, grains of one orientation thin more in the through-

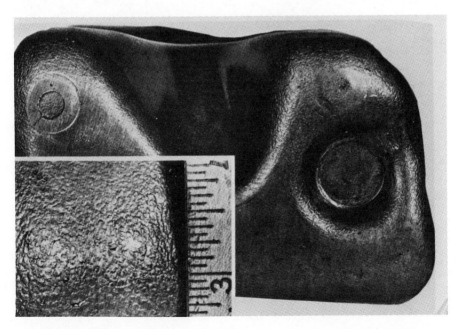

Figure 16-1   Example of orange peel. Courtesy of American Iron and Steel Institute.

thickness direction than those of the other orientation, so a surface roughness appears as relatively large ridges parallel to the rolling direction.

If the grain size is so large that there are less than about five grains across the sheet thickness, forming limits can be appreciably lowered.† With so few grains to the cross section, statistically some regions will be weaker than others. This condition is aggravated if the grain size is not uniform and there are clusters of large grains. In some aluminum alloys, clusters of grains of similar orientations can cause ridging and lowered forming limits.

Stretcher strains, another type of defect, are really incomplete Lüders bands formed in sheet material during forming. These are particularly apparent in regions where the strain is very low, as shown in Fig. 16-2. Stretcher strains can occur in materials that have a pronounced yield point or which develop one by strain aging, (e.g., low-carbon steels). The use of non-aging steels or roller leveling the sheet prior to forming can eliminate this problem. The negative strain-rate sensitivity of some non-ferrous alloys also causes stretcher strains.

The surface roughness of hot-rolled steels precludes their use where surface finish is important.

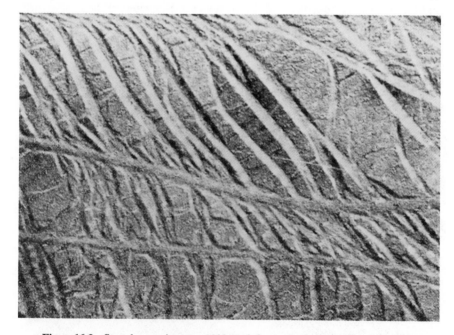

Figure 16-2   Stretcher strains on a 1008 steel sheet stretched past the yield point (7/8 size). From *Metals Handbook, 8th Ed., Vol. 7,* (1972), American Society for Metals, Metals Park, Ohio p. 14.

†D. V. Wilson, W. T. Roberts, and P. M. B. Rodrigues, *Met. Trans.*, 12A, (1981), pp. 1595–1602.

Low-carbon steels (%C ~0.06) are usually finished by cold rolling, except in heavy gauges (> ~0.06 in.), and marketed after a recrystallization anneal at 500–600°C. Most of these steels can be classified as *rimmed* or *killed*. A rimmed steel is one which is not deoxidized before solidification. During freezing of the ingot, dissolved carbon and oxygen react to give a violent evolution of CO bubbles ("rimming") which stirs the molten metal. This stirring breaks up boundary layers and allows segregation of carbon toward the center, so the resulting ingot and the sheet produced from it have surfaces of very low carbon content. These clean surfaces tend to be free of defects caused by carbide particles. In contrast, killed steels are deoxidized by additions of aluminum or silicon during or just prior to ingot pouring. These elements deoxidize the melt so there is no oxygen available to produce CO and therefore no rimming action. Consequently, solidification is still (hence the term "killed"), and boundary-layer formation prevents surface-to-center segregation.

The tensile stress-strain curve of an annealed low-carbon steel shows a pronounced yield point phenomenon (Fig. 16-3). Loading is essentially elastic until some region yields (point *A*). Then the load suddenly drops to a lower yield stress (point *B*). Continued extension of the specimen occurs by growth of the yielded region at a more or less constant stress level.

During this period there is a fairly sharp boundary or *Lüder's band* between the deformed and undeformed regions. Behind this front, all of the

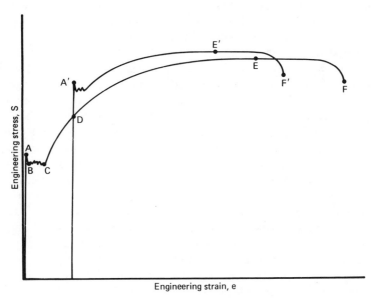

Figure 16-3 Schematic stress-strain curves for a low-carbon steel illustrating strain aging.

metal has been deformed to the same strain. The *Lüder's strain*, or yield point elongation, is typically 1 to 3 %. Only after the Lüders band has traveled the length of the specimen (point C) does strain hardening and uniform straining begin. Finally, the tensile strength (point E) is reached, the specimen necks, and finally fails (point F).

If the specimen were unloaded at some point, D, and then reloaded within a short time, the stress-strain curve would follow the original curve through D, E, and F. However, if the steel were allowed to *strain age* between unloading and reloading, a new yield point, A', would develop, the tensile strength would be raised to E', and the uniform elongation reduced. From the standpoint of formability, such strain aging is undesirable because of the reduction of the strain-hardening exponent and uniform elongation, as well as causing stretcher strains.

Segregation of interstitially dissolved nitrogen (and to a lesser extent, carbon) to dislocations is responsible for the yield point effect. A higher stress is required initially to move the dislocations away from the interstitials than to continue motion of dislocations once they are free. However, the diffusion rate of nitrogen to dislocations is sufficiently fast so that rimmed steel will strain age in several weeks or months at room temperature or in several days at 100°F. Aluminum-killed steels are much more resistant to strain aging because aluminum combines with the nitrogen. Strain aging, however, will occur at higher temperatures, as in an automobile paint baking cycle (~ 200°C). At this stage, strain aging is desirable, since all forming is finished and a return of a yield point increases dent resistance. In commercial practice, the yield point is eliminated by *roller leveling*, which is a bending and unbending over rolls, or by *temper rolling*, which is a very light reduction of ~ 0.5%.

Rimmed steels are somewhat less expensive than killed steels. The cost differential is largely due to less scrap in ingot casting because entrapped CO bubbles reduce shrinkage. With killed steels, a relatively large shrinkage cavity forms in the riser at the top of the ingot, which must be removed before rolling.

Another significant difference between rimmed and killed steels is the degree of normal anisotropy. For killed steels, $\bar{R}$ is usually between 1.4 and 1.8, while for rimmed steels $\bar{R}$ of 1.0 to 1.4 is more common, but there can be considerable variations within each grade depending upon rolling and annealing practice. Higher values of $\bar{R}$ and n accompany increased grain size as shown in Fig. 16-4. *Batch annealing*, in which large coils of sheet are slowly heated to the annealing temperatures over several days, results in larger grain sizes than continuous annealing, which is very rapid. However, grain sizes larger than ASTM Grain Size No. 7 are usually avoided because of excessive orange peel. It should be recognized that there are substantial variations from the trend lines of Fig. 16-4. Such variations occur from plant to plant, within a coil from center to edge, and along the length of a coil.

Hot-rolled sheet is less expensive than cold-rolled sheet, but suffers from poor surface finish and gauge control. Typically $\bar{R} \simeq 1.0$.

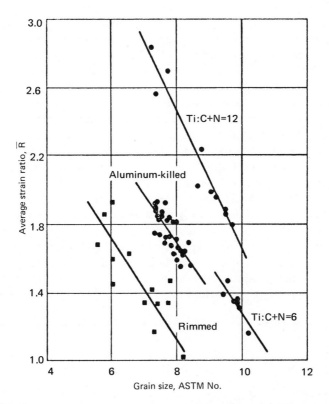

Figure 16-4   Increase of average strain ratio with increasing grain size for low carbon steels. From D. J. Blickwede, *Metals Prog.* (April 1969), p. 90.

There are special grades of low-carbon steel. One of these is *interstitial-free* steel, in which titanium additions have reduced the free carbon and nitrogen so much ($<0.01\%$) that there is no yield point even after annealing. The forming behavior is very good because of low yield strengths, high n-values, and high $\bar{R}$ (often 2 or greater). The disadvantages are higher cost and the lack of strain aging during the paint-bake cycle, which results in weaker parts. *Enameling iron* is used for appliances which have a porcelain-enamel finish. For good adherence, the surface carbon content must be very low. This is achieved by a wet hydrogen anneal of the cold-rolled sheets which allows the reaction $H_2O + \underline{C} \longrightarrow H_2 + CO$.

16-4 HIGHER STRENGTH STEELS

In recent years there has been an increased utilization of higher strength steels with low carbon content. These are used in automobiles to achieve weight savings and hence increased gas mileage, and in pipeline construction. Steels with higher than normal manganese contents and rephosphorized and renitrogenized grades have somewhat higher strengths than the usual low-carbon steels, but have lower formability. Still higher strengths are available in the HSLA (high-strength low-alloy) grades. In these steels, higher strengths are achieved by rapid cooling to achieve a very fine ferrite grain size and a dispersion hardening by carbides or carbonitrides. These steels contain phosphorous, nitrogen, silicon and manganese for solid solution strengthening and vanadium, niobium, or titanium to form fine carbides (or carbonitrides). Most of the HSLA steels are rapidly cooled, just after hot rolling, by a water spray on the runout table, but cold-rolled HSLA steels are produced in some gauges by rapid cooling on a continuous annealing line.

For carbon and HSLA steels, there is a general decrease of the strain-hardening exponent with increased yield strength, as shown in Fig. 16-5. Similarly, the strain-rate exponent tends to fall with strength level. Consequently, greater forming problems are encountered at the higher strength levels.

Because of this there is current interest in dual-phase steels where both high strengths and good formability are required. The dual-phase structure is produced by quenching low-carbon steels from the $\alpha + \gamma$ phase region to form

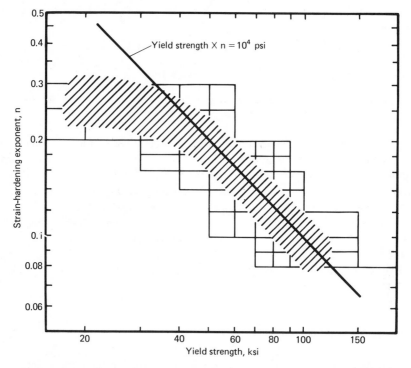

Figure 16-5 Correlation of strain-hardening exponent and yield strength for steels. Adapted from S. P. Keeler and W. G. Brazier in *Micro Alloying 75*, Union Carbide, N.Y. (1977), pp. 517–30.

a microstructure of martensite islands surrounded by ferrite. The relative amount of the two phases depends upon the carbon content and temperature from which they are quenched. These steels have a relatively low yield strength, but because of a high n-value they work harden rapidly to strength levels comparable with the HSLA grades. The high n-value aids in formability, but unfortunately the n-value tends to drop at high strains, so the formability is not as good as that of the low-carbon grades.

Although formability of low-carbon steels is limited by localized necking, the higher strength grades may fail by cracks initiated at elongated inclusions. This is a particularly severe problem at sheared edges when sharp bends are made with the bend axis parallel to the rolling direction or when sheared edges parallel to the transverse direction are elongated during forming. Sulfide inclusions, elongated along the rolling direction, become easy fracture sites. This problem can be alleviated by inclusion shape control. See Chap. 11.

16-5 STAINLESS STEELS

Formable stainless steels fall into two classes: austenitic stainless steels, which are fcc and typically contain $\sim 18\%$ Cr, $\sim 8\%$ Ni, and $<0.15\%$ C; and ferritic stainless steels, which are bcc and typically contain 12 to 18% Cr, $<0.12\%$ C, and no nickel. The ferritic grades are less expensive and find wide application for decorative trim. Austenitic grades are used for high temperature applications and where superior corrosion resistance is required.

The mechanical properties of the ferritic grades are similar to those of low-carbon steels except that the yield and tensile strengths are somewhat higher. Ridging problems noted in Sec. 16-2 can be controlled by annealing at high temperatures. Austenitic grades work harden very rapidly and consequently have high uniform elongation in tension. The austenite in some of these grades is metastable, so low-carbon martensite may be formed during cold working, which partially accounts for the high strain-hardening exponent.

Austenitic stainless steels, like most fcc metals, have low $\bar{R}$-values.

16-6 ALUMINUM ALLOYS

A wide range of aluminum alloys is used in sheet-forming applications. Forming characteristics vary with grades and tempers, but in general, aluminum alloys are not as formable as low-carbon steel. Where high strength is not important (e.g., cooking utensils) commercially pure aluminum (alloy 1100) or alloys containing about 1% Mn (e.g., 3003 and 3004) may be formed in the annealed condition. Strain hardening exponents of ~ 0.25 lead to good formability. However, many applications of aluminum alloys require much higher strengths, particularly when aluminum is being substituted for steel to achieve weight savings. Considerable solid-solution strengthening can be achieved with 2 to

5% Mg (5xxx series). In the annealed condition, the 5182 alloy has an n-value of about 0.3 and formability is good. However, the formation of stretcher strains often limits application where appearance is important. Still higher strengths may be achieved by using these alloys in the cold-rolled condition, but the lower strain-hardening capacity limits formability where stretching occurs. High strength levels can also be obtained in age-hardenable alloys containing Cu (2xxx series) or Mg and Si (6xxx series). The alloy 2036 has reasonable formability in the naturally aged condition and is used for automotive stampings. Canoe halves are formed from alloy 6061 in the solution-treated condition and then are artificially aged.

In general, aluminum alloys have $\bar{R}$-values less than one and the strain-rate sensitivity is very low (even negative) at room temperature. Often, excessive failures result when aluminum alloy sheets are formed with tools that have successfully formed steel, but when the tools are redesigned, parts can be made from aluminum alloys.

16-7 COPPER AND BRASS

Both annealed copper and brass strain harden rapidly; brass having somewhat higher strain hardening exponents, n (0.45 to 0.6 for brass versus 0.35 to 0.5 for copper). For both materials, higher annealing temperatures result in higher values of n, but also in larger grain sizes which induce stronger orange peel effects. Both have a very low rate sensitivity and the normal anisotropy is usually low ($\bar{R} = 0.6$ to 0.9). If copper sheet is produced by annealing after cold rolling to reductions of 80% or greater, a cube texture is likely to be formed in which the $\langle 100 \rangle$ directions are aligned with the rolling, transverse, and through-thickness directions. In this case the sheets will have very low values of $\bar{R}$ and high values of ΔR and, in cupping, large ears will be formed at 0 and 90° to the rolling direction. In brass parts, residual stresses cause a sensitivity to stress-corrosion cracking or season cracking in atmospheres containing ammonia. Splitting occurs along grain boundaries perpendicular to tensile stresses. Even a very low concentration of ammonia in the air from household cleansers or the decomposition of urine may cause cracking after several years. The susceptibility to stress-corrosion cracking can be prevented by giving formed parts a stress-relief anneal (500°F).

16-8 HEXAGONAL CLOSE-PACKED ALLOYS

Zinc alloys tend to have high m-values, low strain-hardening exponents, and low $\bar{R}$-values. In contrast, α-titanium alloys typically have $\bar{R}$-values of 3 to 5 and can be drawn into very deep cups. Sheet forming of most magnesium alloys is generally done at temperatures of 200°C or higher since the ductility at room temperature is very limited.

16-9 PRODUCT UNIFORMITY

Within a coil, the thickness and properties may vary from place to place. Variations along the length of the coil can result from inhomogeneity in the ingot and from temperature variations during hot rolling. There may also be differences between edge and center. These property variations reflect differences in grain size, texture, and composition. Lack of uniformity is one reason why, for a given coil, some stampings fail and others do not. If there are significant differences in thickness and structure over short distances (e.g., within a blank to be stamped), uniform elongation and formability will be reduced for the reasons discussed in Chap. 4.

PROBLEMS

16-1. The table below, from "Making, Shaping, and Treating of Steels," (Pittsburgh: United States Steel Co., 1971), p. 1127, gives combinations of aging times and temperatures to achieve equal amounts of strain aging in low-carbon steels.

  (a) From a plot of ln t versus $1/T$, determine the apparent activation energy for strain aging.

  (b) Can you explain why the slope changes between 0 and 21°C? (Think about how data were probably obtained.)

Aging Times at Several Temperatures Required to Produce Approximately Equal Aging Effects

0°C	21°C	100°C	120°C	150°C
1 yr	6 mo	4 h	1 h	10 min
6 mo	3 mo	2 h	30 min	5 min
3 mo	6 wk	1 h	15 min	2.5 min
1 mo	2 wk	20 min	5 min	
1 wk	4 d	5 min		
3 d	36 h	2 min		

16-2. Aluminum-killed drawing-quality steel sheets cost 5 to 10% more than rimming steels. One stamping plant saved millions of dollars annually by lowering the specifications on certain parts from AK to rimmed steel.

  (a) If you were in a stamping plant, how would you select the parts for similar lowering of specifications?

  (b) Would substituting of rimmed for killed steel result in an inferior product? Explain.

16-3. With low-carbon steels, both the n- and R-values can be raised by using higher annealing temperatures after cold rolling. Why isn't this practice more widely used?

16-4. The substitution of HSLA steels for low-carbon grades to achieve weight saving in automobiles is based on their higher yield strengths.

  (a) What implications does such substitution have for the elastic stiffness of components?

(b) In view of corrosion on cars, what implication does such substitution have for component life? (*Hint:* the corrosion resistance of HSLA steels is almost identical to that of low-carbon steels.)

16-5. The table below lists properties of several sheet materials at the temperature at which they will be deformed.

Properties

Material	E $(10^6$ psi)	Y.S. $(10^3$ psi)	R_0	R_{45}	R_{90}	n	m
A	30	32	1.9	1.2	2.0	0.25	0.03
B	30	35	1.2	1.0	1.2	0.22	0.03
C	10.5	25	0.7	0.6	0.7	0.22	0.002
D	16.5	20	0.6	0.9	0.6	0.50	0.001
E	10	1	1.0	1.0	1.0	0.00	0.60

(a) Which of the materials would have the highest LDR in cupping? On what basis did you choose this material?

(b) Which material would show the greatest amount of earing during cupping? On what basis did you choose this material?

(c) Which material would have the greatest uniform elongation in a tension test? On what basis did you choose this material?

(d) Which material would have the greatest total elongation in a tension test? On what basis did you choose this material?

(e) Excluding material E (because of its low yield strength), which material could be formed into the deepest cup by a hemispherical punch acting on a clamped sheet (no "drawing")? On what basis did you select this material?

Index

Index

A

Alligatoring, 232
Aluminum alloys, 319-20
Anisotropy, 263
    crystallographic basis, 263-65
    effect on limiting drawing ratio,
      275-80
    strain ratio (R-value), 264-66
    yield criteria, 266-72

B

Bauschinger effect, 29
Bending, 250
    shapes and tubes, 257-58
    springback and residual stresses,
      251-58
    with superimposed tension,
      255-56
Bridgman correction factor, 60-61
Bulge test, 63-64
Bulging, 193-96

C

Centerline cracking (also called
    "chevron" cracks and "cuppy
    cores"), 228-30
Compression:
    direct or axisymmetric, 61
      average pressure, 125-26
      barreling, 61
      slab analysis, 125-26
      test, 61-62
      upper-bound analysis, 156
    plane-strain:
      average pressure, 122-23
      slab analysis, 120-24
      slip-line field analysis,
        194-98
      test, 64-65
      upper-bound analysis,
        155-60
Copper and brass, 320
Cracking, edge (*See also* Fracture):
    in rolling, 136-38, 238-39
    in sheet forming, 305